STUDENT UNIT GUIDE

CCEA® AS

UNIT 1

Biology

Molecules and Cells

John Campton

Philip Allan Updates, an imprint of Hodder Education, an Hachette UK company, Market Place, Deddington, Oxfordshire OX15 0SE

Orders
Bookpoint Ltd, 130 Milton Park, Abingdon, Oxfordshire, OX14 4SB
tel: 01235 827827
fax: 01235 400401
e-mail: education@bookpoint.co.uk
Lines are open 9.00 a.m.–5.00 p.m., Monday to Saturday, with a 24-hour message answering service. You can also order through the Philip Allan Updates website: www.philipallan.co.uk

ISBN 978-0-340-99193-0

First printed 2010
Impression number 5 4
Year 2014 2013 2012 2011

This guide has been written specifically to support students preparing for the CCEA® AS Biology Unit 1 examination. The content has been neither approved nor endorsed by CCEA and remains the sole responsibility of the author.

CCEA is a registered trademark, the use of which has been licensed to Hodder but does not constitute endorsement by CCEA.

Printed in India

Hachette UK's policy is to use papers that are natural, renewable and recyclable products and made from wood grown in sustainable forests. The logging and manufacturing processes are expected to conform to the environmental regulations of the country of origin.

Contents

Introduction

■ ■ ■

Content Guidance

■ ■ ■

Questions and Answers

Introduction

About this guide

The aim of this guide is to help you prepare for the AS Unit 1 examination for CCEA® biology. It also offers support to students studying A2 biology, since topics at A2 rely on an understanding of AS material.

This guide has three sections:
- **Introduction** — this provides guidance on the CCEA® specification and offers suggestions on improving your study/revision skills and examination technique.
- **Content Guidance** — this summarises the specification content of AS Unit 1.
- **Questions and Answers** — this provides two exemplar papers for you to try. There are answers written by two candidates and examiner's comments on the candidates' performances and how they might have been improved.

Try to adopt the suggestions given in this introduction about how to study. This will affect your performance throughout the course — it takes time to learn how to study at this level.

The Content Guidance should be used as a study aid as you meet each topic, for end-of-topic tests, and during your final revision. There are seven topics and at the end of each topic there is a list of the practical work with which you are expected to be familiar.

The Questions and Answers section will be particularly useful during your final revision. It presents a range of question styles that you will encounter in the AS Unit 1 exam, and the candidates' answers and examiner's comments will help with your examination technique.

The specification

You should have your own copy of the CCEA® biology specification. This is available from **www.ccea.org.uk**.

AS biology

The AS course is made up of three units. Units 1 and 2 are assessed by examination papers; Unit 3 is assessed internally and has a lower weighting.

AS Unit	Title	Weighting	Availability
1	Molecules and Cells	40% of AS	January and summer
2	Organisms and Biodiversity	40% of AS	January and summer
3	Assessment of Practical Skills	20% of AS	Summer

This guide covers AS Unit 1 and the second book in the series covers AS Unit 2. AS Unit 3 is based on the practical work (i.e. coursework) that you will do in your biology classes.

Assessment objectives

AS biology is not just about remembering facts. Examiners need to assess your *skills* — your ability to do things and work things out.

Examinations in AS biology test three different assessment objectives (AOs). AO1 is about remembering the biological facts and concepts covered by the unit. AO2 is about being able to use the facts and concepts in new situations. AO3 is called How Biology Works. It emphasises that biology, as a science, develops through testing hypotheses. It involves an understanding of practical procedures and the analysis of results to determine whether a hypothesis has been supported or disproved.

The following table gives a breakdown of the approximate number of marks awarded to each AO in the examination.

Assessment objective	Description	Marks
AO1	Knowledge and understanding of biology and of How Biology Works	32 (42.5%)
AO2	Application of knowledge and understanding of biology and of How Biology Works	32 (42.5%)
AO3	How Biology Works	11 (15%)

The AS Unit 1 paper

The AS Unit 1 examination lasts 1 hour 30 minutes and is worth 75 marks. The paper consists of about nine questions; the mark allocation per question is from 3 to 15 marks. There are two sections. In Section A all the questions are structured; in Section B there is a single question, which may be presented in several parts, and which should be answered in continuous prose.

Questions towards the start of the paper and the initial parts of questions tend to assess straightforward 'knowledge and understanding' (AO1). There will also be questions that present information in new contexts and may test your skills in analysing and evaluating data (AO2).

AO3 may be assessed by questions that ask you to describe experimental procedures. You may also be asked to demonstrate graphical or drawing skills, since these are important aspects of biology. Graphs are used to illustrate quantities; drawings illustrate the qualities of a feature. There are other skills that you may be asked to demonstrate — for example, organisation of raw data into a table.

You are expected to use good English and accurate scientific terminology in all your answers. Preparing a glossary of terms used in each topic should aid this. Quality of written communication is assessed throughout the paper and is specifically awarded a maximum of 2 marks in Section B.

Study and revision skills

Students who achieve good grades have good study strategies. This section of the introduction provides advice and guidance on how these might be achieved.

Revision is an ongoing process

Revision should not be just something that you do before an exam. It should be continual throughout the course. Work consistently and complete each task as the teacher sets it. Use study periods in school to develop your understanding, *not* for homework. Study thoroughly for tests. Try to find time at the weekend to go over that week's work. If you keep going over topics then you won't panic with the intensive revision required at exam time.

Active learning is best

Just reading through your notes or a chapter of a textbook is not a particularly effective way to revise. You simply learn material to later forget it. In order to develop a deeper understanding, you have to use more than just the 'reading' centre of the brain — you should make your brain *do* something with the material. It is for this reason that you must write your own notes. You should also try to do things in different ways, for example:

- a series of bullet points
- a flow diagram
- an annotated diagram
- a spider diagram
- a prose account

Use different methods for each topic. A spider diagram on enzymes would include reference to theoretical aspects, the effects of temperature and pH, the effect of inhibitors and the effect of immobilisation. A series of annotated graphs would be a good way to revise enzyme properties. Compiling a glossary of terms to do with enzymes will improve your understanding of how each term is defined. An essay on enzymes will test your understanding of the entire topic and give you practice for the Section B question.

Developing your understanding

You should learn to develop your understanding so that you can apply it. Use different texts and the internet. These will present the information in different ways, so that your brain can perceive it from different perspectives. It is only when you understand a topic fully that you will be able to deal with questions that set the topic in a new context. If you have problems with a difficult concept, ask your teacher to explain it in a different way. Teachers are happy to help, as long as you have worked at the topic yourself. Working with another student may also help. Remember too that AS biology is a step up from GCSE. There will be difficult concepts and so you must persevere.

Organise your notes

You will have accumulated a large quantity of notes from your teacher and, more importantly, those that you have made yourself. You should organise your notes under headings and sub-headings and construct a summary of the key points. The Content Guidance section will help you do this. It is essential that you keep this information in an organised manner. Gather all your notes, divide them into sections on each topic and keep them in a file for each unit. This will make it easier both when you add additional notes and during your final revision.

Planning your revision

While revision is an ongoing process, you will have to undertake intensive revision in the weeks before the exam. Make out a revision schedule taking into account all the topics you have to cover and the time available. This can seem alarming, but if you break down the total amount into smaller portions then it will become achievable. Make sure that each part of the unit gets its fair share of your attention and allocate more time to difficult areas. Try to leave time in your schedule to practise past questions.

Keeping your concentration

Some students get lost in their work and can concentrate for hours. Some find it difficult to concentrate for longer than an hour. You should do whatever is appropriate for you, but try to revise in a quiet place with no distractions. You should take breaks, which could be organised as rewards — for example, a favourite television programme.

Vary your revision and keep it active

In the final revision it is easy to revert to just reading through your notes. Try to keep it active by summarising what you have. This keeps your brain processing the information. Try to vary what you do. You should test your understanding by practising past questions, such as in the exemplar papers in the Question and Answer section of this guide. You should practise skills such as calculations and also write essays on the various topics in the unit. Scanning can be an effective revision technique: read the first sentence of each paragraph (the rest of the paragraph generally only provides elaboration); read sentences with key words (often in bold); and, in particular, study diagrams since these often summarise important information. You might also find the website at **http://highered.mcgraw-hill.com/sites/dl/free/0072437316/120060/ravenanimation.html** useful for animations of many key processes.

7

The examination

Before the day of the examination it is important that you are well prepared. You should have all the implements that you will need: two black pens and two pencils, your calculator plus spare batteries, a ruler and an eraser. Try to get a good night's sleep. When you enter the exam hall, tell yourself that you really understand this unit — *be positive*. You will always know more than you think you do. A few nerves are good and will help you stay more alert during the exam.

Time

You have 90 minutes to answer questions worth a total of 75 marks. That gives you over 1 minute per mark, so there is some preparation time and you should have some time at the end to go over your answers. At the start of the exam it can be beneficial to spend some time looking through the paper and scanning the questions, especially the question in Section B. This will allow you to think of relevant points while answering other questions. If you get stuck, make a note of the question number and move on — you can come back later. You are advised to spend 20 minutes on Section B. Try to keep to this, but leave some time for writing a plan of the information you want to include. At the end of the exam go over the paper, including those parts about which you were unsure, correcting any mistakes or filling in any gaps. It is also beneficial to double-check calculations to make sure that you haven't made any silly mistakes.

Read the questions carefully

This sounds obvious, but you can lose many marks by not doing so. There are two aspects:

- You must understand the command terms used in the question, i.e. the word at the start of the question. Appendix 1 of the specification and a guide in the biology microsite on **www.ccea.org.uk** explain these terms. The two terms that are most commonly misinterpreted are 'describe' and 'explain' (there is a further note about this in the Content Guidance section).
- The stem of a question may provide information needed to answer the question. Don't ignore this information. You might wish to highlight or underline these parts of the question stem. Think about how this information can help you to construct or focus on a relevant answer.

Depth and length of answer

The examiners give you guidance about how much you need to write:

- **The number of marks**. In general, the number of marks indicates the number of points that you should provide — for a question worth 4 marks you need to give more points than for one worth 2 marks. For questions testing straightforward recall, you may need to provide more points than there are marks — for example, three points for 2 marks. However, this should be obvious from the number of spaces. The number of marks is the most important guideline with respect to the depth of answer required.

- **The number of lines**. In general, there will be two lines for each marking point. However, examiners expect you to keep your answer relevant and precise, so occasionally there may be only one line. In Section B, the number of lines allocated is generous — a good answer will not necessarily use them all.
- **The recommended time**. You are advised to spend 20 minutes on Section B. Try to keep to this. It is possible to write a longer 'essay' but you would be providing more points than there are marks available.

Quality of written communication

The ability to organise thoughts, express ideas clearly and to make use of the appropriate terminology is an important aspect of biology. In Section A questions, credit may be restricted if communication is unclear. Where quality of written communication (QWC) is assessed in Section A, the mark schemes for questions will contain specific statements. These statements will relate to the clear expression of concepts (e.g. the active site of an enzyme has a complementary shape to that of its substrate, but not the same shape), correct biological terminology (e.g. use of the term water potential, rather than water concentration) and correct spelling where there are close alternatives (e.g. mitosis and meiosis). Note that mistakes in spelling are not generally penalised as long as the examiner knows what you are trying to say. With respect to clarity in answers, a common problem is to use the word 'it' in such a way that the examiner can't be certain what 'it' refers to — so try to avoid using 'it' in answers. In Section B, there is a maximum of 2 marks available for QWC. The examiners want to see well-linked sentences which present relationships and do not just list features.

Content Guidance

This section summarises what you need to know and understand for the AS Unit 1 examination paper. It is divided into seven topics:

- **Biological molecules** — water, ions, carbohydrates, lipids and proteins
- **Enzymes** — the theory of enzyme action, properties of enzymes, enzyme cofactors and coenzymes, enzyme inhibitors and immobilised enzymes
- **Nucleic acids** — the structure of nucleotides and nucleic acids, DNA replication, DNA as the genetic code and DNA technology
- **Cells and viruses** — microscopy, the eukaryotic cell (animal, plant and fungal), the prokaryotic cell and viruses (bacteriophage and HIV)
- **Membrane structure and function** — the structure of the cell-surface membrane and the movement of substances in and out of the cell
- **The cell cycle, mitosis and meiosis** — the processes of mitosis and meiosis, and the genetic consequences of mitosis and meiosis
- **Tissues and organs** — the ileum and the leaf

At various points within the section there are examiner's tips. These offer guidance on how to avoid difficulties which often occur in examinations.

At the end of most topics there is a list of practical work with which you should be familiar.

Biological molecules

The chemical composition of living organisms

All living organisms are composed of atoms. The atoms occur as parts of small molecules (e.g. glyceraldehyde, $C_3H_6O_3$), large molecules (e.g. haemoglobin, $C_{3032}H_{4816}O_{872}N_{780}S_8Fe_4$), polyatomic ions (e.g. phosphate, PO_4^{3-}) and single ions (e.g. potassium ion, K^+). Only 20 different types of atom (of the 92 stable elements) occur in living organisms. The elements present in the largest proportions are carbon (C), hydrogen (H), nitrogen (N), oxygen (O), phosphorus (P) and sulphur (S). These atoms have low atomic mass and combine with one another to form molecules held together by strong **covalent bonds**. Consequently, living organisms are both light and strong. The remaining atoms occur as charged ions (e.g. Ca^{2+}, Na^+, K^+, Cl^-, Fe^{2+}).

The **water content** of living organisms ranges from 50% to 95%. Water is composed of hydrogen and oxygen; its formula is H_2O. Water is a liquid whereas other substances of similar molecular mass are gases (e.g. CH_4, NH_3, O_2 and CO_2). The reason for this is that water is a **dipolar** molecule (the oxygen is slightly negative, δ– and the hydrogen is slightly positive, δ+), so neighbouring water molecules are linked by **hydrogen bonds** — weak bonds between the oxygen on one water molecule and a hydrogen on another (see Figure 1).

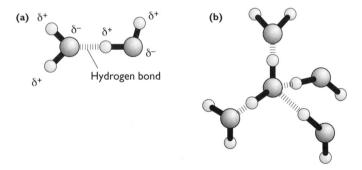

Figure 1 (a) The charges on water molecules; (b) a cluster of water molecules

The dry mass of living organisms is in the form of carbohydrates, lipids, proteins and nucleic acids.

Water is a good **solvent** capable of dissolving a wide range of chemical substances. This includes all ions and molecules with charged groups. Ions are charged and are surrounded by shells of water; water clusters around the charged groups of glucose and amino acids. This is shown in Figure 2.

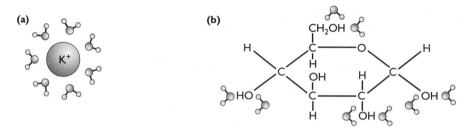

**Figure 2 (a) K⁺ with a shell of water molecules;
(b) glucose with clusters of water molecules**

The biochemical reactions within the cell are carried out in solution. (Since the water content of seeds and spores is as low as 10% their biochemical activity is suspended until they become rehydrated.) Water is also used to transport nutrients and waste substances.

The hydrogen bonding of water molecules is known as **cohesion**. This is an important property in the flow of a continuous column of water (and dissolved nutrients) through the xylem vessels of plants. You will study this in AS Unit 2.

Water also has an important role in temperature regulation since evaporation of water from a surface cools it down. The energy required to break the hydrogen bonding in liquid water is known as the **latent heat of evaporation**.

Since **ions** are soluble in water this is the way in which living organisms absorb certain elements. The atoms contained within the ion have specific roles (see Table 1).

Table 1 The role of ions and their atoms as components of biologically important molecules

Ion	Chemical symbol	Role in biological molecule
Nitrate	NO_3^-	Nitrogen in the amino group of amino acids produced in plants
Sulphate	SO_4^{2-}	Sulphur in the R group of the amino acid cysteine
Phosphate	PO_4^{3-}	In a range of important molecules: adenosine triphosphate, ATP; nucleotides and so nucleic acids; phospholipids
Calcium	Ca^{2+}	In calcium pectate, which contributes to the middle lamella of plant cell walls; in calcium phosphate in the bone of vertebrate animals
Magnesium	Mg^{2+}	In the chlorophyll molecule
Iron	Fe^{3+}	In the haemoglobin molecule

Since ions and polar molecules (e.g. glucose and amino acids) are charged and have shells or clusters of water, this influences how they pass through cell-surface membranes (see p. 54).

Carbohydrates

Carbohydrates contain carbon, hydrogen and oxygen. They have the general formula $C_x(H_2O)_y$. The simplest carbohydrates are single sugars (**monosaccharides**) with the formula $(CH_2O)_n$, where n can vary from 3 to 9. The important types of monosaccharide are **trioses** ($n = 3$), **pentoses** ($n = 5$) and **hexoses** ($n = 6$). Two hexose sugars bond together to form a **disaccharide** (double sugar) via a **condensation** reaction (a chemical reaction in which two molecules are joined together and one molecule of water is removed). **Polysaccharides** are formed when many hexose sugars are linked by condensation reactions. Disaccharides and polysaccharides release hexose sugars via **hydrolysis**.

Important monosaccharides include:
- the pentose sugars, **ribose** and **deoxyribose** — these are constituents of nucleotides that form the nucleic acids RNA and DNA
- the hexose sugars, α-**glucose**, β-**glucose** and **fructose** (see Figure 3)

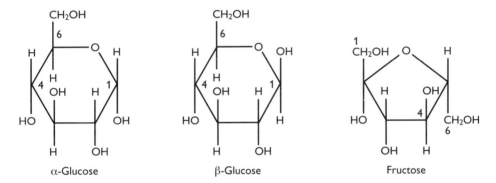

Figure 3 α-glucose, β-glucose and fructose

The carbon 1, carbon 4 and carbon 6 positions are indicated in the glucose molecules shown above. These are important since it is at these positions that different glucose molecules bond together. The subtle but important distinction between α-glucose and β-glucose is that the –H and –OH groups at the carbon 1 position are reversed. This means that the two types of glucose bond slightly differently.

The hexose sugars can bond together to produce disaccharides:
- α-glucose and fructose form **sucrose** — sucrose (cane sugar) is the form in which carbohydrates are transported in plants
- α-glucose and α-glucose form **maltose** — the product of starch digestion

The formation of maltose is a **condensation reaction** since water is removed in the process. The bond formed is an **α-1,4-glycosidic bond** (see Figure 4). Note that the two α-glucose molecules are not in the same plane. The breaking of a glycosidic bond is a **hydrolysis reaction**. In this case, maltose would be hydrolysed to its constituent α-glucose molecules. The formation and hydrolysis of maltose is shown in Figure 4.

α-Glucose

CH₂OH

H   H

4   1

HO   OH

CH₂OH

H   H

4   1

HO   OH

α-Glucose

Condensation   **Hydrolysis**

H_2O H_2O

Maltose

CH₂OH

H

4   1

HO

CH₂OH

H

4

1

OH

O

α-1,4 glycosidic bond

Note that the
bond causes
the α-glucoses
to lie at different
angles

Figure 4 The formation and hydrolysis of maltose

Many α-glucose molecules can bond to produce polysaccharides. The simplest is **amylose**, a constituent of starch in plants (see Figure 5). Only α-1,4-glycosidic bonds are involved and a helical structure results since the α-glucoses join at slightly different angles to each other.

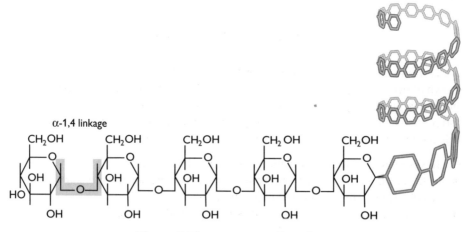

α-1,4 linkage

CH₂OH   CH₂OH   CH₂OH   CH₂ OH   CH₂OH

O   O   O   O   O

OH   OH   OH   OH   OH

HO

O   O   O   O

OH   OH   OH   OH   OH

Figure 5 The structure of amylose

Two other polysaccharides are produced when **α-1,6-glycosidic bonds** are added at regular intervals. In **amylopectin**, also a constituent of starch, α-1,6-bonds occur every 24 to 30 glucose units; in **glycogen**, found in the liver and muscle cells of mammals α-1,6-bonds occur every 8 to 10 glucose units. The structure of a branched polysaccharide is shown in Figure 6.

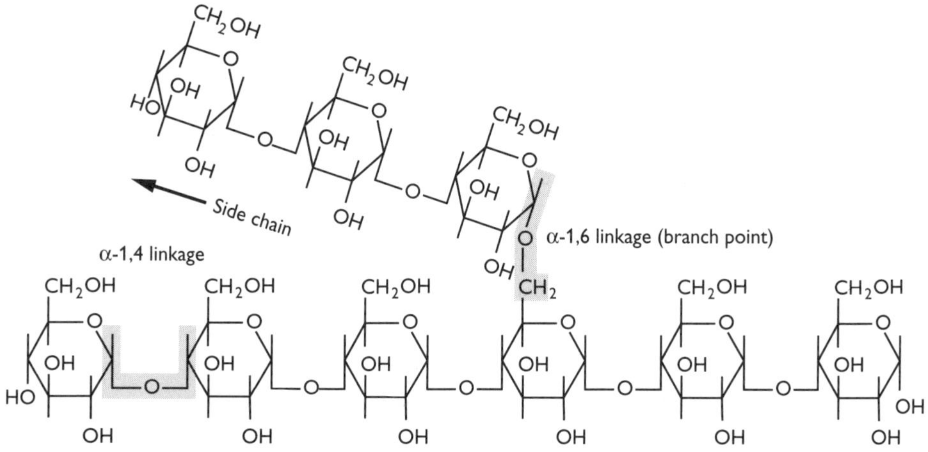

Figure 6 The structure of a branched polysaccharide: amylopectin and glycogen

Glucose is an important respiratory substrate. The polysaccharides amylose and amylopectin in starch, and glycogen, are the means by which glucose is stored in organisms (so they are regarded as energy stores). They are well adapted as storage molecules because:

- many glucose molecules can be stored in a cell (amylose molecules are compact)
- they are readily hydrolysed to release glucose molecules (especially amylopectin and glycogen, which are branched and so have many terminal ends for the hydrolytic enzymes to act)
- they are insoluble and so cannot move out of cells
- they are osmotically inert (only soluble substances affect the water potential of a cell — see pp. 52–53)

Many β-glucose molecules bond to produce the polysaccharide **cellulose**. Due to the way in which the β-glucose molecules are orientated cellulose molecules are straight chains. Adjacent cellulose molecules are linked by hydrogen bonds and grouped to produce **microfibrils**. Each microfibril consists of hundreds of cellulose molecules and is a structure of immense tensile strength. Cellulose microfibrils mesh to form the **cell wall** of the plant cell and prevent the cell from bursting in dilute solutions. The structure of cellulose is shown in Figure 7.

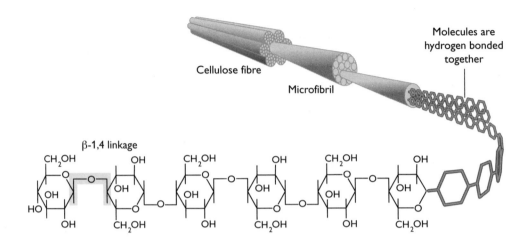

Figure 7 The structure of cellulose

Some polysaccharides are compared in Table 2.

Table 2 A summary of different polysaccharides

Polysaccharide	Monomer	Glycosidic bond(s)	Shape of polymer	Location	Function
Amylose	α-glucose	α-1,4 bonds only	Unbranched, helical molecule	Starch grains in living plant cells	Store of glucose (energy)
Amylopectin	α-glucose	α-1,4 and 1,6 bonds	Branched, helical molecule	Starch grains in living plant cells	Store of glucose (energy)
Glycogen	α-glucose	α-1,4 and 1,6 bonds	Branched, helical molecule	Granules in liver and muscle cells of mammals	Store of glucose (energy)
Cellulose	β-glucose	β-1,4 bonds only	Straight chains cross-linked to parallel chains	Cell walls of plant cells	Structural support to plant cell

Lipids

Lipids contain mostly carbon and hydrogen with a few atoms of oxygen; they may contain other types of atom. They are macromolecules, but they are not polymers. They are a diverse group structurally, the common feature being their insolubility in water. They can be extracted from cells by organic solvents. Examples include triglycerides (fats and oils), phospholipids, steroids (e.g. cholesterol) and waxes.

Fatty acids are an important constituent of triglycerides and phospholipids. They are essentially long hydrocarbon chains with a carboxylic acid group at one end. They can vary according to:
- the length of the hydrocarbon chain
- whether the hydrocarbon chain contains double bonds — a hydrocarbon with double bonds is described as **unsaturated** in comparison with chains with only single bonds that are **saturated** with hydrogen (see Figure 8)

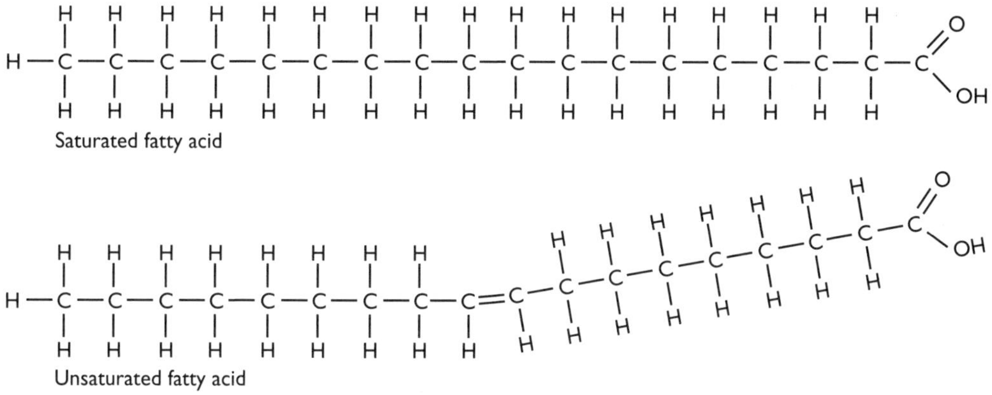

Figure 8 is described below the images.

Saturated fatty acid

Unsaturated fatty acid

Figure 8 A saturated fatty acid and an unsaturated
fatty acid with one double bond

Triglycerides consist of a **glycerol** molecule bonded by condensation reactions to three fatty acids. **Ester bonds** are formed. The constituent molecules of a triglyceride are released by hydrolysis reactions. The formation and hydrolysis of a triglyceride are shown in Figure 9.

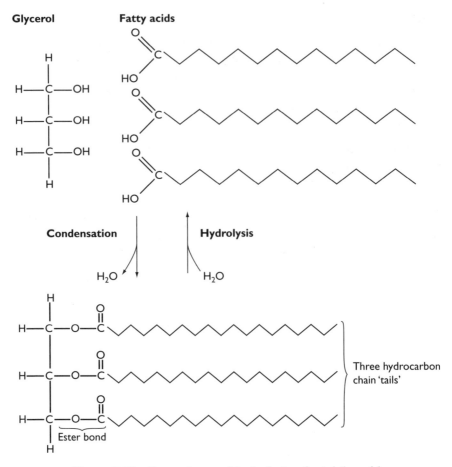

Figure 9 The formation and hydrolysis of a triglyceride

Triglycerides with unsaturated hydrocarbon chains (or with shorter chains) have lower boiling points than those with saturated hydrocarbon chains. Therefore:

- triglycerides with unsaturated hydrocarbon chains tend to be liquid at room temperatures — oils
- triglycerides with saturated hydrocarbon chains are solid — fats

Oils tend to be found in plants while fats occur in animals.

Like polysaccharides, triglycerides represent energy stores (particularly the constituent fatty acids). They represent an efficient means of storing energy since gram-for-gram they release more energy that carbohydrate. As a mass-efficient means of storing energy lipids are found in seeds (e.g. linseed oil), migratory birds (e.g. ducks) and in the camel's hump. Fats are also important in providing:

- a thermal insulating layer in mammals, since they are poor heat conductors
- buoyancy in marine mammals such as dolphins and whales

• a cushioning layer around and, therefore, protection to internal organs such as the kidneys

Fatty acids may only be respired aerobically.

A **phospholipid** consists of a glycerol molecule, two fatty acid residues and a phosphate group (see Figure 10). The phosphate causes the glycerol end (the 'head') to be polarised and, therefore, soluble in water (**hydrophilic** or 'water-loving'); the long hydrocarbon chains (the 'tails') are non-polar and insoluble in water (**hydrophobic** or 'water-hating').

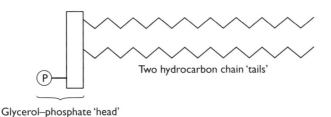

Two hydrocarbon chain 'tails'

Glycerol–phosphate 'head'

Figure 10 A phospholipid

In an aqueous environment, phospholipids automatically form bilayers. The phospholipid bilayer represents the basis of membrane structure in the cell (see pp. 50–51).

The steroid **cholesterol** is essentially a molecule with a hydrocarbon chain and four carbon-based rings. It is found in cell membranes and, since it is hydrophobic, is found among the hydrocarbon chains of the phospholipid bilayer. A number of steroid hormones are synthesised from cholesterol, including the sex hormones oestrogen and testosterone.

Proteins

Proteins make up about two-thirds of the total dry mass of a cell. They contain carbon, hydrogen, oxygen, nitrogen and usually sulphur. Proteins are chains of amino acids. Since there are 20 different amino acids, which can be arranged in many different sequences, a huge variety of proteins is possible. Proteins have a highly organised structure with up to four levels of organisation. The overall shape is precise and integral to the function of the protein in the cell.

Amino acids consist of a carbon atom with four groups attached:
• an amino group
• a carboxylic acid group
• a hydrogen atom
• a residue (R-group)

It is the residue that differs to form the 20 different naturally occurring amino acids. Some of the residues carry a charge and so may be involved in hydrogen bonding, some are hydrophobic and a few contain sulphur (e.g. cysteine). The general structure of an amino acid is shown in Figure 11.

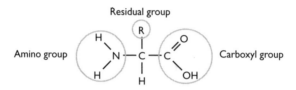

Figure 11 The general structure of an amino acid

Amino acids can bond together to form a **dipeptide**. A condensation reaction is involved and the amino acids are linked by a **peptide bond**. A hydrolysis reaction breaks the dipeptide down to release the two amino acids. The formation and hydrolysis of a dipeptide are shown in Figure 12.

Figure 12 The formation and hydrolysis of a dipeptide

Many amino acids are peptide bonded together to form a **polypeptide**. The **primary structure** of a polypeptide is the sequence of amino acids in the chain. The polypeptide has an amino group at one end and a carboxyl group at the other.

The **secondary structure** is either an α-helix or a β-pleated sheet. The structures are held in place by hydrogen bonds between peptide links in adjacent parts of the chain.

Globular proteins have a **tertiary structure**. The polypeptide folds over on itself in a precise way to produce a specific three-dimensional shape. This is due to interaction between the free R-groups of the amino acids. Different R-groups produce specific links with each other: hydrogen bonds between polar R groups; hydrophobic interactions between non-polar R-groups; ionic bonds between ionised R-groups; disulphide bonds between the sulphur-containing R-groups of cysteine residues. A different order of amino acids means that the R-group interactions are different and so a different three-dimensional shape is generated. Fibrous proteins lack a tertiary structure.

Some proteins consist of two or more polypeptide chains bonded together. This is the **quaternary structure**.

The levels of organisation in a protein molecule are shown in Figure 13.

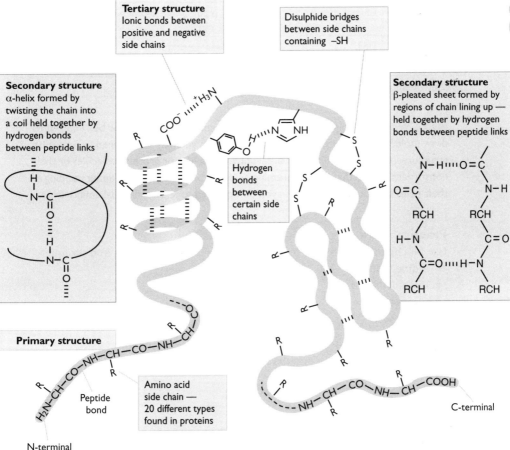

Figure 13 The different levels of structure in a protein molecule

Haemoglobin is a **globular** protein found in large quantities in red blood cells. Each molecule consists of four polypeptides: two α-chains and two β-chains. Each polypeptide has an iron-containing haem group attached. Haemoglobin is important in the transport of oxygen in animals. An oxygen molecule can associate with each haem to form oxyhaemoglobin.

Collagen is a **fibrous** protein. Each molecule consists of three identical poly-peptides coiled round each other and held together by hydrogen bonds. Collagen molecules are bonded together to form the strong fibres found in the skin, tendons and ligaments.

Conjugated proteins have a non-protein part attached. The non-protein component is called a **prosthetic** group. Some conjugated proteins are given in Table 3.

Table 3 Some important conjugated proteins

Name	Prosthetic group	Location
Glycoprotein	Carbohydrate	Mucin (component of saliva); cell-surface membrane
Lipoprotein	Lipid	Membrane structure
Nucleoprotein	Nucleic acid	Chromosome structure; ribosome structure
Haemoglobin	Haem (iron-containing)	Red blood cells

Tip With the exception of a generalised amino acid, you will not be required to draw molecular structures from scratch. However, you may be required to recognise structures, or complete diagrams of carbohydrates, lipid or protein molecules or their constituents. Note that individual amino acids are only shown to aid understanding of their roles in protein structure — you do not need to learn individual amino acids.

Practical work

Use biochemical tests to detect the presence of carbohydrates and proteins:
- iodine test
- Benedict's test
- Clinistix
- Biuret test

Carry out paper chromatography of amino acids:
- preparation, running and development of the chromatogram
- calculation of R_f values

Enzymes

The chemical reactions of an organism are collectively called **metabolism**. Metabolic reactions include:
- **catabolism** — 'breakdown' reactions
- **anabolism** — 'build-up' reactions

Enzymes catalyse metabolic reactions – there is one type of enzyme for each reaction.

The theory of enzyme action

In order to catalyse a reaction, the enzyme and substrate must first collide to form an **enzyme–substrate complex**. Catalysis then takes place on the enzyme surface, according to the equation below:

E + S		ES		EP		E + P
Enzyme + substrate	⟶	Enzyme–substrate complex	⟶	Enzyme–product complex	⟶	Enzyme + product

The enzyme is unchanged at the end of the reaction (and the same products are formed whether the reaction is catalysed or uncatalysed).

The reaction takes place on a particular part of the enzyme molecule called the **active site**. In an anabolic reaction, the substrate molecules are orientated in such a way on the active site as to allow bonding between them. In a catabolic reaction, the formation of the active site round the substrate assists the breaking of a particular bond. In both cases the enzyme functions as a catalyst by effectively lowering the **activation energy** required for the reaction to take place. This reduction in activation energy is illustrated in Figure 14.

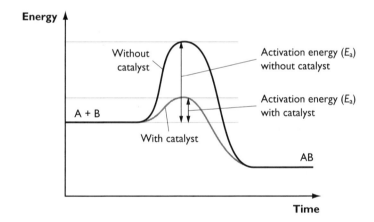

Figure 14 The effect of a catalyst on the activation energy of a reaction

Enzymes are **specific**. For any one type of reaction there is a particular enzyme required for catalysis — if there are 1000 different kinds of reaction in a cell, then the cell contains 1000 different enzymes.

There are two models to explain how enzymes work:
- the **lock-and-key** model (see Figure 15)
- the **induced-fit** model (see Figure 16)

The lock-and-key model of enzyme action proposes that the active site of an enzyme has a complementary shape (like a lock) into which the substrate molecule (the key) fits exactly, to form the enzyme–substrate complex. The induced-fit model suggests that initially, the shape of the active site is not quite complementary to that of the substrate, but as the substrate begins to bind, the active site changes shape and 'moulds' itself around the substrate molecule (like a glove fitting round a hand). The induced-fit model is considered more useful because it better explains the way in which activation energy is reduced in catabolic reactions.

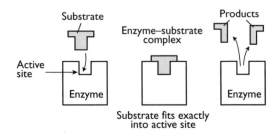

Figure 15 The lock-and-key model of enzyme action

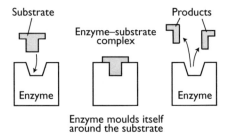

Figure 16 The induced-fit model of enzyme action

Properties of enzymes

A number of external factors influence the activity of enzymes and, therefore, the rate of biological reactions. These include substrate concentration, enzyme concentration, temperature and pH.

There are two key points to remember when explaining these influences:
- Enzyme molecules need to collide with substrate molecules, so factors that influence the chance of collision, such as substrate concentration and temperature, influence the rate of reaction.
- Enzymes are globular proteins with a precise tertiary structure, so factors that influence protein shape, such as high temperature and pH, influence the rate of reaction.

The effect of substrate concentration on enzyme activity

The influence of substrate concentration is summarised in Figure 17.

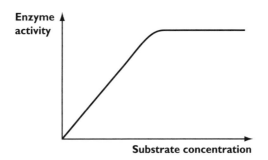

Figure 17 The effect of substrate concentration on enzyme activity

- At low substrate concentrations, an increase in concentration increases enzyme activity. This is because a greater concentration of substrate molecules increases the chances of collision with enzyme molecules. Therefore, more enzyme–substrate complexes are formed.
- At high substrate concentrations, an increase in concentration does not cause a further increase in activity. This is because at high substrate concentrations the enzymes are fully employed and so, at any one moment, all the active sites are occupied.

Tip Questions relating to graphs often ask you to 'describe' and/or 'explain' the trends. 'Describe' may seem simple enough, but there are pitfalls. For the plateau part of the graph shown in Figure 17 don't be tempted to write that 'there is no effect' — there is still a high level of activity, it is just that there is no further increase. If the question asks you to 'explain' then you must use your understanding of the biological process to provide reasons for the trends illustrated. In the example above, you would show your understanding of how collision theory relates to enzyme action.

The effect of enzyme concentration

The influence of enzyme concentration is summarised in Figure 18.

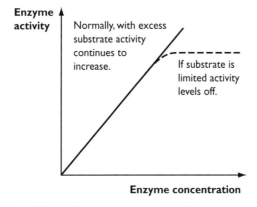

Figure 18 The effect of enzyme concentration on enzyme activity

An increase in enzyme concentration increases the rate of reaction. At high enzyme concentration activity may level off, but only if there is insufficient substrate. This is because an increase in the concentration of enzyme molecules increases the chance of successful collisions with substrate molecules. (There is normally only an incline phase since enzymes are used over-and-over again and so function at very low concentrations.)

The effect of pH on enzyme activity

The influence of pH is summarised in Figure 19.

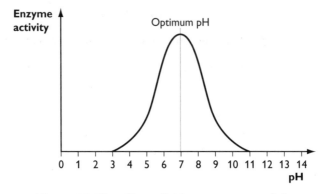

Figure 19 The effect of pH on enzyme activity

Enzyme activity is at a maximum at the optimum pH. An increase or decrease in pH causes a decrease in enzyme activity. This is because the structure of the protein and, therefore, the active site of the enzyme are altered by changes in pH. In particular, ionic bonds in the tertiary structure may be disrupted. So at non-optimal pH, the substrate attaches less readily to the enzyme and there is a specific pH at which the shape of the active site best facilitates the formation of an enzyme–substrate complex.

The effect of temperature on enzyme activity

The influence of temperature is summarised in Figure 20.

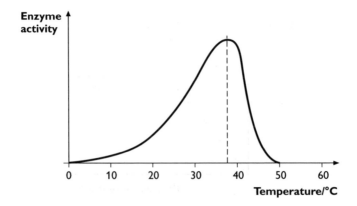

Figure 20 The effect of temperature on enzyme activity

- At low temperatures, an increase in temperature causes an exponential increase in enzyme activity — typically, a 10°C rise in temperature doubles the rate of reaction. This is because an increase in temperature provides more kinetic energy for the collision of enzyme and substrate, so the rate of formation of enzyme–substrate complexes increases.
- At high temperatures (typically above 40°C), an increase in temperature causes a sharp decline in enzyme activity. This is because at higher temperatures, the bonds holding the tertiary structure of the enzyme molecules are broken and so the active site loses its complementary shape for substrate attachment — the enzyme is denatured.

Not all enzymes have an optimum temperature of around 40°C. Enzymes from organisms that live in very cold habitats have a much lower optimum temperature. Enzymes from bacteria that live in hot springs are active at temperatures up to 90°C. For example, the DNA polymerase enzyme used in the polymerase chain reaction (PCR) has an optimum temperature of 80°C and was obtained from the thermophilic bacterium *Thermophilus aquaticus*, found living in hot springs.

Enzymes, cofactors and coenzymes

Some enzymes do not function effectively unless a non-protein **cofactor** is attached. Cofactors include metal ions, such as Mg^{2+}, and organic molecules (**coenzymes**) that are often derivatives of vitamins. Cofactors function either by influencing the shape of an enzyme (to its optimum for substrate attachment) or by participating in the enzymatic reaction (by attaching to one of the products for transfer to another enzyme). Some examples of cofactors and coenzymes are given in Table 4.

Table 4 Examples of cofactors and coenzymes

Enzyme	Cofactor	Role of enzyme
Carbonic anhydrase	Zinc ion (Zn^{2+})	Catalyses the combination of CO_2 with water to form carbonic acid in red blood cells, facilitating the transport of CO_2 in the blood
Cytochrome oxidase — a respiratory enzyme	Copper ion (Cu^{2+})	Combines electrons and hydrogen ions with oxygen in respiration
Enzyme	**Coenzyme**	**Role of enzyme and coenzyme**
Pyruvate decarboxylase— a respiratory enzyme	Coenzyme A	Pyruvate (3-carbon molecule) is broken down to acetate, which is 'picked up' by coenzyme A (forming acetyl CoA), and CO_2, which diffuses out of the cell
Succinate dehydrogenase — a respiratory enzyme	FAD (derived from vitamin B_2, riboflavin)	Hydrogen is removed from succinate and 'picked up' by FAD (to form $FADH_2$)

Tip The examples shown in Table 4 are provided only to illustrate the roles of cofactors and coenzymes. You do not need to learn these — most are involved in respiratory metabolism, which is covered in A2 Unit 2.

Enzyme inhibitors

Enzyme inhibitors are molecules that bind to enzymes and decrease their activity.

- A **competitive inhibitor** closely resembles the structure of the substrate and so competes for the active site, but does not remain there permanently (see Figure 21).

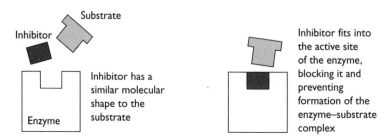

***Figure 21 A competitive inhibitor competes
with the substrate for the active site***

- A **non-competitive inhibitor** binds to the enzyme (not necessarily at the active site) and changes the shape of the active site or blocks it irreversibly so that the substrate can no longer attach.

The effects of both types of inhibitor on enzyme activity at increasing substrate concentration are shown in Figure 22.

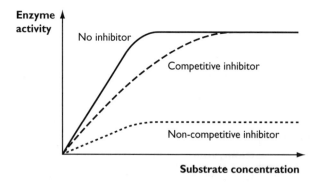

***Figure 22 The effect of substrate concentration on enzyme activity in the
presence of a competitive inhibitor and a non-competitive inhibitor***

Effect of competitive inhibitor: the degree of inhibition depends on the relative concentration of both inhibitor and substrate because each is competing for a place on the active site. The more substrate there is available the more likely it is that a substrate molecule will find an active site. Therefore, if the substrate concentration is increased, the effect of the inhibitor is reduced.

Effect of non-competitive inhibitor: the substrate and the inhibitor are not competing for the same site, so an increase in substrate concentration does not decrease the effect of the inhibitor.

Some examples of enzyme inhibitor are shown in Table 5.

Table 5 Examples of enzyme inhibitors

Enzyme	Inhibitor	Type
Succinate dehydrogenase — a respiratory enzyme	Malonate — similar in structure to succinate (the substrate)	Competitive inhibitor
Cytochrome oxidase — a respiratory enzyme	Potassium cyanide, KCN — combines with the active site	Non-competitive inhibitor (and irreversible)

Immobilised enzymes

An immobilised enzyme is an enzyme that is attached to an inert, insoluble material. Methods of immobilisation include:

- adsorption on glass, alginate beads or matrix — the enzyme is attached to the outside of an inert material
- entrapment — enzyme is trapped inside insoluble beads or microspheres, for example calcium (or sodium) alginate beads
- cross-linkage — the enzyme is bonded covalently to a matrix by a chemical reaction

Enzyme immobilisation facilitates the use of continuous-flow column reactors (see Figure 23).

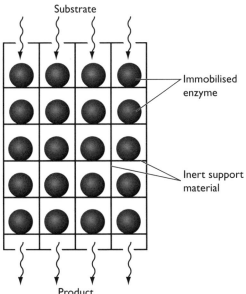

Figure 23 Immobilised enzymes in a continuous-flow column reactor

Immobilisation provides advantages in the commercial use of enzymes:
- The product is enzyme-free.
- The enzyme can be re-used.
- Since the enzyme is supported, its stability is improved. This means that it remains active over a greater range of pH and temperatures (thermostability) than would be the case if the enzyme were in solution.

However, immobilisation can reduce enzyme activity:
- The active site of the immobilised enzyme may be blocked by the support matrix (adsorption).
- Insoluble substances may hinder the arrival of the substrate (entrapment).
- The active site may be altered during the binding process (cross-linkage).

Further, the flow rate through the column influences activity. If it is too slow the reaction is completed early in the column; too fast and not all the substrate will have been engaged in the reaction.

Uses

Immobilised enzymes have a variety of commercial uses. Enzymes can be immobilised onto reagent dip-strips for diagnostic purposes, e.g. the enzymes glucose oxidase and peroxidase on Clinistix.

> **Tip** In the examination, you may be asked to construct a graph. A **graphical-skills question** may ask you to 'plot the results, using an appropriate graphical technique'. You will have decisions to make:
> - What type of graph is most appropriate?
> - What is the most appropriate caption (summarising the relationship shown in the graph)?
> - Which is the independent variable (to be plotted on the x-axis)?
>
> In constructing the graph, you should ensure that the axes are labelled and that any units of measurement are shown; that the points (or bars) are accurately plotted (and joined by short straight lines if that is appropriate); and that if there is more than one line then these are identified (e.g. with labels or a key).
>
> For example, in an experiment using a colorimeter to follow the course of a starch–amylase catalysed reaction the graph shown in Figure 24 might be drawn. Note that here a smooth curve is drawn since a curvilinear process is predicted by theoretical considerations and a curve is indicated by the trend of the points.
>
> You may be asked to describe and explain the trends evident in the graph. In this case, you would note that the rate of starch breakdown is rapid initially and decreases as the reaction progresses — it is an exponential decline curve. The reason for this is that the concentration of starch is high initially, so there are frequent collisions with the amylase molecules but as starch is hydrolysed the concentration is reduced and so fewer collisions occur.

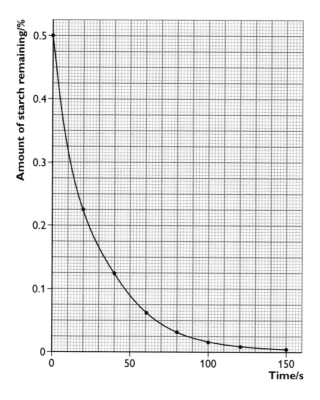

Figure 24 The course of a starch–amylase reaction

Practical work

Carry out experimental investigation of factors affecting enzyme activity:
- effect of temperature, pH, substrate and enzyme concentrations on enzyme activity
- demonstration of enzyme immobilisation
- use of a colorimeter to follow the course of a starch–amylase reaction (or other appropriate reaction)

Nucleic acids

Nucleic acids are macromolecules composed of chains of nucleotides. They carry coding information and are found in all living cells and viruses. There are two forms: **deoxyribonucleic acid (DNA)** and **ribonucleic acid (RNA)**.

The role of DNA is the long-term storage of genetic information:

- It is the means by which genetic information is passed from generation to generation. It is able to do this because it is capable of **self-replication**.
- It acts as a **code**, since it contains the instructions needed to construct other cell components, such as proteins and RNA. The code on DNA is represented by the sequence of bases in the nucleotides.

Lengths of DNA that carry the genetic information for the synthesis of proteins are called **genes**. Genes consist of nucleotides with a specific sequence of bases. The sequence of the bases determines the sequence of amino acids in a polypeptide, coding for the order in which amino acids are brought together.

RNA molecules assist the functioning of DNA. They retrieve information from DNA and direct the synthesis of proteins. RNA molecules act as messengers between DNA in the nucleus and the sites of protein synthesis (the ribosomes) in the cytoplasm. They also form portions of the ribosomes and serve as carrier molecules for amino acids to be used in protein synthesis.

Nucleotide structure

Nucleotides are the subunits of nucleic acids. Each nucleotide (see Figure 25) consists of:

- a **pentose** sugar – **deoxyribose** or **ribose**
- a **nitrogenous base** – from **adenine, guanine, cytosine** and **thymine** (in deoxyribonucleotides only) or **uracil** (in ribonucleotides only)
- a **phosphate** group (attached to the carbon-5 of the sugar)

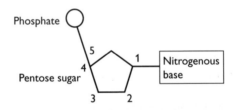

Figure 25 A generalised nucleotide (the numbers indicate the position of carbon atoms in the 5-carbon sugar)

Deoxyribonucleotides and ribonucleotides are compared in Table 6.

Table 6 A comparison of deoxyribonucleotides and ribonucleotides

Feature	Deoxyribonucleotides	Ribonucleotides
Pentose sugar	Deoxyribose	Ribose
Nitrogenous base	Adenine (A), guanine (G), cytosine (C), thymine (T)	Adenine (A), guanine (G), cytosine (C), uracil (U)
Macromolecule formed	DNA	RNA

Nucleic acid structure

Nucleotides join together by condensation reactions forming **phosphodiester bonds** (between the phosphate of one nucleotide and the C3 of the pentose of the other nucleotide) along a 'sugar–phosphate' backbone. The polynucleotide strand formed has a free **5'-end** (with phosphate attached) and a free **3'-end**. This is illustrated in Figure 26.

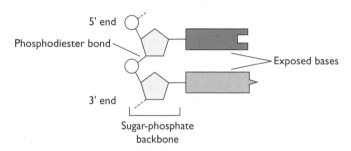

Figure 26 The basic structure of a polynucleotide

A DNA molecule consists of **two anti-parallel strands**. The bases on opposite strands are joined by hydrogen bonds in a precise way: adenine always bonds with thymine; guanine always bonds with cytosine (see Figure 27). This is known as base pairing

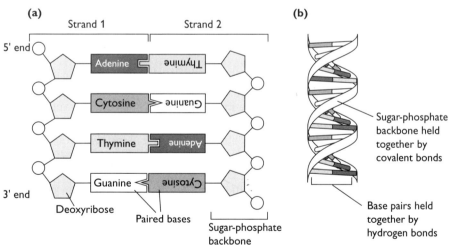

Figure 27 (a) The structure of DNA; and (b) formed into a double helix

The nucleotides join at slightly different angles to each other and so the whole structure forms a **double helix**; the hydrogen bonding between the two strands increases its stability.

RNA molecules are single stranded and are much shorter than DNA molecules — DNA may be millions of nucleotides long, RNA usually consists of a few hundred nucleotides. There are three forms of RNA:

- **Messenger RNA (mRNA)** carries the code for the synthesis of a polypeptide from the DNA in the nucleus to a ribosome where the polypeptide is assembled.
- **Transfer RNA (tRNA)** carries the amino acids to the ribosome to be used in polypeptide synthesis. tRNA is single stranded and folded into a clover-leaf shape, with hydrogen bonding within the folds.
- **Ribosomal RNA (rRNA)** forms part of the structure of ribosomes. These organelles are the sites at which polypeptides are assembled.

DNA and RNA are compared in Table 7.

Table 7 A comparison of DNA and RNA

Feature	DNA	RNA
Subunits	Deoxyribonucleotides (contains deoxyribose and thymine)	Ribonucleotides (contains ribose and uracil)
Length	Very long	Relatively short
Types	One (though nucleotide sequences differ)	Three: mRNA; tRNA; rRNA
Strands	Double stranded	Single stranded (though tRNA and rRNA have folds that are bonded)
Base pairing	A with T and G with C	No base pairing (except joining folds within tRNA and rRNA)

DNA replication

Since DNA is the genetic code for the synthesis and development of whole organisms, it must be copied exactly from one generation to the next. This is achieved by self-replication, using a **semi-conservative mechanism** in which each strand acts as a template for the synthesis of a new strand. Each new DNA molecule contains one of the original strands in addition to a new strand (hence the name semi-conservative).

The sequence of events in DNA replication is as follows:

(1) The enzyme DNA helicase breaks the hydrogen bonds holding the base pairs together and 'unzips' part of the DNA double helix, revealing two strands.

(2) The enzyme DNA polymerase moves along each strand, which acts as a template for the synthesis of a new strand.

(3) DNA polymerase catalyses the joining of free deoxyribonucleotides to each of the exposed original strands, according to base pairing rules, so that new complementary strands form.

(4) The process of unzipping and joining new nucleotides continues along the whole length of the DNA molecule.

Each DNA molecule so formed is identical to the other and to the original DNA (and contains one strand of the original).

DNA replication is illustrated in Figure 28.

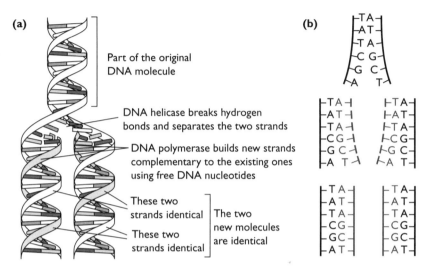

(a)

Part of the original DNA molecule

DNA helicase breaks hydrogen bonds and separates the two strands

DNA polymerase builds new strands complementary to the existing ones using free DNA nucleotides

These two strands identical

These two strands identical

The two new molecules are identical

(b)

Figure 28 (a) DNA replication; (b) semi-conservative replication of DNA

The evidence for semi-conservative replication: Meselson and Stahl's experiment

Experimental evidence that DNA replicates semi-conservatively came from a classic experiment devised by Matthew Meselson and Franklin Stahl in 1958. They grew bacteria (*Escherichia coli*) in a medium in which nitrogen was supplied (in ammonium ions) in the form of the heavy, but non-radioactive isotope, ^{15}N. Consequently, the DNA of the bacteria became entirely heavy.

These bacteria were then transferred to a medium containing the normal (light) isotope, ^{14}N. Immediately before changing the medium and then at intervals corresponding to successive generations, samples of bacteria were removed and the DNA was extracted.

Analysis of the extracted DNA involved density-gradient centrifugation, a technique that separates molecules of different molecular masses; heavier molecules are deposited at a lower level in the centrifuge tube. The results were as predicted by the semi-conservative hypothesis. Immediately before the change of medium, the DNA occupied a single band corresponding to 'heavy' DNA. After one generation the DNA was of 'intermediate' density, being concentrated in a single band a little higher up the tube. This is to be expected if all the DNA molecules consist of one heavy strand and one light strand. After two generations there were two bands, with 50% of the DNA 'intermediate' and 50% 'light'. The results and their interpretation are shown in Figure 29.

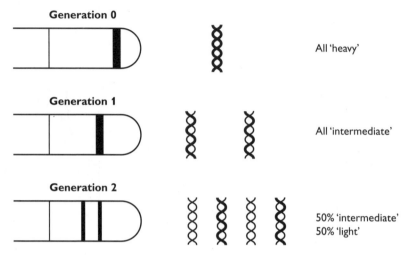

Figure 29 Meselson and Stahl's experiment

The genetic code

A **gene** is a length of DNA that codes for a protein (polypeptide). The protein may be structural (e.g. collagen) or functional (e.g. membrane proteins and enzymes). Since proteins control the activities of organisms, the genes ultimately determine their characteristics.

Humans possess about 30000 genes, which represents only 10% of the total DNA. Between the genes there are extensive non-coding regions of DNA, the function of which is uncertain.

The order of bases on one strand of the length of DNA that forms the gene determines the order by which amino acids are sequenced during the formation of a polypeptide. The bases are 'read' in triplets (triplet code). Each triplet of bases acts as a code for a specific amino acid (just as '...' is the Morse code for the letter 's') — for example:

Bases in the DNA strand:	ATG	GCT	GAA	TGT
Amino acid sequence:	methionine	alanine	glutamate	cysteine

The process by which proteins are synthesised is dealt with in detail in A2 Unit 2.

DNA technology

The nucleotide sequence of our DNA determines the essence of our being. The nucleotide sequences of our genes (the coding regions) determine the proteins that we produce. The nucleotide sequences of the non-coding regions, while not well understood, are highly variable, which makes them valuable in determining how closely related two individuals might be. The ability to study the base sequences of DNA allows a number of interesting questions to be tackled:

- Does a particular gene differ from person to person?
- Is an allele (a particular form of a gene) associated with a hereditary disease?
- Are two persons closely related; are two DNA samples the same?
- How much does the DNA of individuals within a population vary; how much genetic diversity is there within a particular species?
- How much does the DNA of different species vary; how can this be used to unravel the taxonomic relationships among species?

Determining the similarities and differences in the nucleotide sequences of DNA samples is possible because of the development of a number of tools. For example, one genetic marker site relies on the use of **restriction endonuclease** enzymes, obtained from certain bacteria. The recognition of specific nucleotide sequences in different DNA fragments relies on the use of **DNA probes**; **DNA profiling** (**fingerprinting**) is used to compare the DNA of individuals. If the initial sample for analysis does not contain sufficient DNA, then the amount can be amplified using the **polymerase chain reaction** (**PCR**).

Restriction endonuclease enzymes

Restriction endonuclease enzymes cut DNA at specific nucleotide (base) sequences. Bacteria produce these enzymes to counter attack by viruses (bacteriophages). They do this by cutting bacteriophage DNA into smaller, non-infectious fragments.

There are many different restriction enzymes produced by different species of bacteria. Each enzyme cuts DNA at a specific base sequence — the recognition (or restriction) site (see Table 8). The enzymes may make staggered cuts in the DNA, commonly called **'sticky' ends**.

Table 8 Some examples of restriction endonuclease enzymes

Restriction endonuclease	Bacterial origin	Recognition site
EcoRI	E. coli	G A A T T C C T T A A G
HindIII	H. influenzae	A A G C T T T T C G A A
BamHI	B. amyloliquefaciens	G G A T C C C C T A G G

The enzyme cuts the DNA producing a particular number of fragment lengths depending on the number of recognition sites in the DNA — if there are four sites then five fragment lengths are produced.

Restriction endonuclease enzymes have been used to establish restriction fragment length polymorphisms (see below) and may be used in DNA profiling. Their use in obtaining DNA segments containing a particular gene is dealt with in A2 Unit 2.

Genetic markers

A genetic marker is a nucleotide sequence that is variable within a population and can, therefore, be used to measure differences between individuals. Three genetic markers are described:

- **Restriction fragment length polymorphisms** (RFLPs — pronounced riflips). The different array of fragment lengths produced when a specific restriction endonuclease enzyme is used to cut different DNA molecules. The variation is due to differences in the number of recognition sites between individuals.
- **Microsatellite repeat sequences** (MRSs). Short runs of simple two-, three-, or four-base sequences found within the non-coding regions of the DNA, which are therefore highly variable. The number of repeats of these microsatellites (e.g. CCTA) varies between individuals and the pattern of MRSs within the DNA of an individual is unique. It is MRSs that are analysed in DNA profiling (fingerprinting).
- **Single nucleotide polymorphisms** (SNPs — pronounced snips). These are differences in single nucleotides among samples of DNA molecules. SNPs within genes are particularly interesting since they may indicate the cause of a genetic disease. For example, a single nucleotide change in the gene that codes for the β-chain of haemoglobin causes sickle-cell anaemia. In this disease, the red blood cells are distorted and the haemoglobin is less effective at carrying oxygen.

The polymerase chain reaction (PCR)

The polymerase chain reaction (PCR) mimics the natural process of DNA replication. It can generate billions of copies of a DNA sample within a few hours. It is also possible to select particular sections of DNA to replicate – for example, sections containing microsatellite repeat sequences (in DNA profiling).

The process requires:

- a DNA sample that includes the selected region for replication
- the synthesis of primers — short strands of DNA (of about 20 nucleotides) that are complementary to the sequence at the start of each strand of the region to be amplified
- the enzyme DNA polymerase, extracted from thermophilic bacteria and which is therefore thermostable
- free deoxyribonucleotides

The sequence of events in PCR (see Figure 30) is as follows:

(1) The DNA to be amplified (copied many times) is heated to 95°C. This breaks the hydrogen bonds and separates the two strands.

(2) The mixture is cooled to 53°C, which allows the primers to bind to the start of each strand of the selected DNA region. The primers prevent the DNA strands rejoining and act as signals to the polymerase enzymes to start adding nucleotides.

(3) The mixture is heated to 73°C and the thermostable polymerase enzyme copies each strand, starting at the primers.

(4) The process is repeated and with each cycle the number of DNA molecules is doubled — a chain of 20 cycles would produce millions of copies of the original DNA.

The technique is used by forensic scientists and archaeologists to study small samples of DNA, and may be used in DNA profiling.

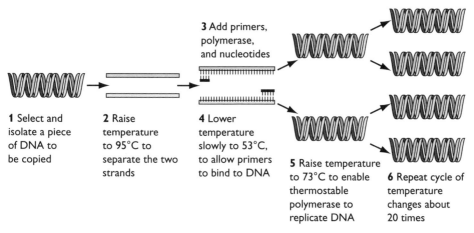

1 Select and isolate a piece of DNA to be copied

2 Raise temperature to 95°C to separate the two strands

3 Add primers, polymerase, and nucleotides

4 Lower temperature slowly to 53°C, to allow primers to bind to DNA

5 Raise temperature to 73°C to enable thermostable polymerase to replicate DNA

6 Repeat cycle of temperature changes about 20 times

Figure 30 Stages in the polymerase chain reaction (PCR)

The DNA probe

A DNA probe is used to locate a particular section of DNA. It consists of a short length of single-stranded DNA with a specific nucleotide (base) sequence. The probe binds by base pairing to a complementary region of single-stranded target DNA. In order to be able to detect the probe and the target DNA to which it is attached, the probe is either radioactive or fluorescent. Detection takes place using an X-ray film for radioactive probes or a laser scanner for a fluorescent probe.

DNA profiling (fingerprinting)

The DNA of the genes that code for the proteins of an individual does not show much variation. This is because differences to the nucleotide sequences would be mutations and could result in non-functional proteins. However, the vast amount of non-coding ('junk') DNA that exists between the genes is highly variable. Within the non-coding regions, **microsatellite repeat sequences** are found. A microsatellite (e.g. CCTA) may be repeated between five and 15 times.

DNA profiling uses MRSs that are very similar between closely related individuals but so variable that unrelated individuals are extremely unlikely to have the same MRSs.

One method of DNA profiling makes *use of restriction endonuclease to cut sections of DNA containing the MRSs*:

- The DNA is treated with a specific restriction endonuclease to cut the DNA into fragments either side of the MRSs. If there are a large number of repeats then the fragment will be long; if there are few repeats small fragments will be produced.

- The fragments of DNA are separated by gel electrophoresis on the basis of size — smaller fragments move further.
- The bands are heated to make them single stranded, transferred (blotted) to a nylon membrane (Southern blotting) and washed with DNA probes that are complementary to the microsatellite region.
- This is repeated using several probes for different MRSs and the resulting fragments are detected.
- Only those DNA fragments that have bound to the labelled probe show up — the resulting pattern of bands is called a DNA fingerprint and looks rather like a bar code.

This technique is illustrated in Figure 31.

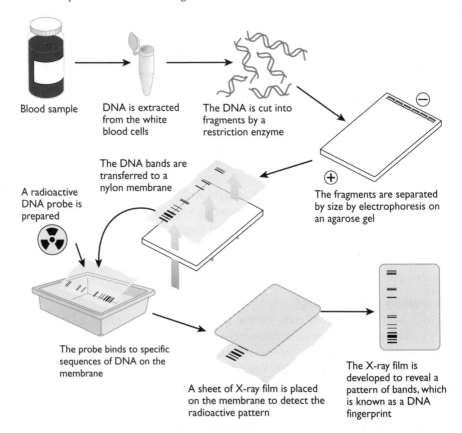

Blood sample

DNA is extracted from the white blood cells

The DNA is cut into fragments by a restriction enzyme

The DNA bands are transferred to a nylon membrane

A radioactive DNA probe is prepared

The fragments are separated by size by electrophoresis on an agarose gel

The probe binds to specific sequences of DNA on the membrane

A sheet of X-ray film is placed on the membrane to detect the radioactive pattern

The X-ray film is developed to reveal a pattern of bands, which is known as a DNA fingerprint

Figure 31 Stages in preparing a DNA profile (genetic fingerprint) using restriction endonuclease

Another method (presently used by the UK Forensic Science Service) makes *use of the polymerase chain reaction to amplify the MRSs*:

- Fluorescent DNA primers that will attach next to the region containing the MRSs are synthesised.

- PCR is used to replicate large numbers of DNA fragments containing the MRSs. This allows minute quantities of source material to be analysed.
- The fragments of DNA produced by PCR are separated using gel electrophoresis.
- This is completed for ten different microsatellites, each four bases in length, with an additional primer for determining gender.
- The position of the DNA fragments is revealed as a pattern of fluorescent bands due to the fluorescent tags on the DNA primers flanking the microsatellite regions. The bands are detected using a laser scanner.
- The results are displayed in a graph of fluorescence against fragment size — the DNA profile.

DNA profiling is a powerful tool in forensic science, in settling paternity disputes, in establishing kinship, and in studying the genetic diversity of species and the evolutionary relationship between taxonomic groups.

Cells and viruses

The cell is the structural unit of all living organisms. Many living organisms consist of just a single cell, while others are composed of many cells. These multicellular organisms possess cells that are adapted to perform specialised functions (e.g. mesophyll cells in a leaf for photosynthesis and epithelial cells in the ileum for absorption). Nevertheless, there are many functions that all cells undertake (e.g. ATP synthesis, protein synthesis) and so they possess common features within their internal structure.

There are two categories of cell: **prokaryotic**; and **eukaryotic**. The prokaryotic cells lack the degree of 'compartmentalisation' that eukaryotic cells possess — they lack a nucleus and other membrane-bound organelles. Bacteria are prokaryotes. Eukaryotes include animals, plants and fungi.

Viruses are not cells and are not themselves living. They have an intimate relationship with, and reproduce in, living cells.

Microscopy and cell ultrastructure

Two types of microscope are used in the study of cells: the **light microscope** and the **electron microscope**. Both **magnify** the fine structure of an object. However, **resolution** is even more important. This is the ability to discriminate fine detail so that two neighbouring points are seen as separate, rather than as a larger blur. The electron microscope has much greater resolving power than the light microscope because electrons have a shorter wavelength than light. The light microscope has its advantages, one of which is that living processes, such as mitosis, can be viewed. The interior of an electron microscope is a vacuum and so specimens must be dead. These two types of microscope are compared in Table 9.

**Table 9 A comparison of the light microscope
and the transmission electron microscope**

Light microscope	Transmission electron microscope
Uses light: wavelength 450–700 nm	Uses electrons: wavelength 0.01 nm
Light refracted by glass lenses	Electron beams refracted by electromagnetic lenses
Low resolution: 200 nm	High resolution: 0.1 nm
Low magnification: ×1500 maximum	High magnification: ×1 000 000 maximum
Image formed on the retina of the eye or recorded on photographic film	Image formed on a fluorescent screen or recorded on photographic film
Limited in cellular detail that is revealed	Limited to dead specimens and by the likelihood of artefacts (deviations from the 'real' appearance as a result of the treatment of the specimen in preparation for microscopy)
Advantage: can be used to view living cells	Advantage: produces high resolution images of cells and organelles

The **scanning electron microscope** (SEM) is similar to the **transmission electron microscope** (TEM) except that the specimen is coated in a film of gold and electrons are reflected off the surface to create a three-dimensional effect image.

> **Tip** In this unit you might be required to calculate the magnification of a photograph or electronmicrograph, or the true (actual) size of an organelle.
>
> $$\text{magnification} = \frac{\text{size of image}}{\text{true size}}$$
>
> $$\text{true size} = \frac{\text{size of image}}{\text{magnification}}$$
>
> All measurements must be in the same units. The conversion factors are as follows:

	× 1000		× 1000		× 1000	
metres (m)	⇌ ÷ 1000	millimetres (mm)	⇌ ÷ 1000	micrometres (µm)	⇌ ÷ 1000	nanometres (nm)

Remember that mathematical calculations must make sense — a mitochondrion cannot be the equivalent of a metre in length!

The eukaryotic cell

The appearance of a cell seen through an electron microscope is its **ultrastructure**. The ultrastructure of a generalised animal cell is shown in Figure 32.

Eukaryotes also include plants and fungi. Their cells have some distinctive features (see Figure 33 — cells not drawn to scale).

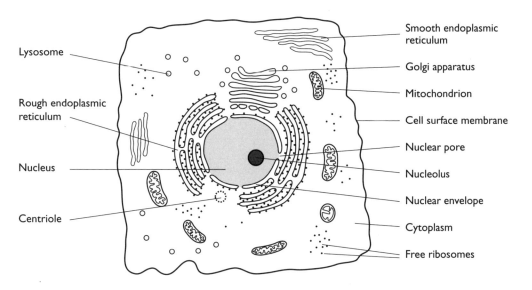

Smooth endoplasmic reticulum

Lysosome

Golgi apparatus

Mitochondrion

Rough endoplasmic reticulum

Cell surface membrane

Nuclear pore

Nucleus

Nucleolus

Nuclear envelope

Centriole

Cytoplasm

Free ribosomes

Figure 32 The ultrastructure of an animal cell

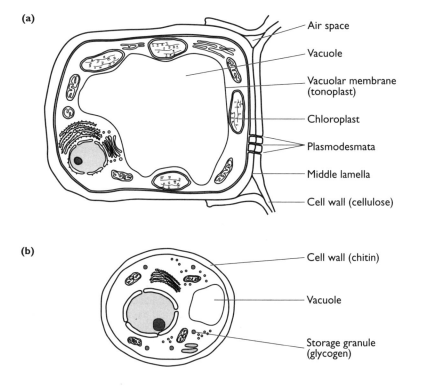

(a)

Air space

Vacuole

Vacuolar membrane (tonoplast)

Chloroplast

Plasmodesmata

Middle lamella

Cell wall (cellulose)

(b)

Cell wall (chitin)

Vacuole

Storage granule (glycogen)

Figure 33 Distinctive features of (a) a plant cell; (b) a fungal cell

Both plant and fungal cells have cell walls, though made of different materials. The cell wall prevents the cell bursting when in dilute solution.

Animal, plant and fungal cells are compared in Table 10.

Table 10 A comparison of animal, plant and fungal cells

Animal cell	Plant cell	Fungal cell
No cell wall	Cellulose cell wall	Chitin cell wall
No chloroplasts	Chloroplasts	No chloroplasts
Glycogen granules (carbohydrate (energy) store)	Starch grains (carbohydrate (energy) store)	Glycogen granules (carbohydrate (energy) store)
Lysosomes	No lysosomes	Lysosomes
No permanent vacuole	Large central vacuole	Vacuole
Centrioles	No centrioles (except mosses and ferns)	No centrioles (except one group)
No plasmodesmata	Plasmodesmata	No plasmodesmata

The structures that perform particular functions within a cell are called **organelles** (see Table 11). Some of these are surrounded by membranes — membrane-bound organelles.

Table 11 The structure and function of eukaryotic cell organelles

Organelle	Structure	Function
Nucleus	Largest organelle (10–30 µm) enclosed within an envelope (double membrane); contains chromosomal DNA which may be extended (euchromatin) or condensed (heterochromatin); perforated envelope (possesses pores); contains one or several nucleoli (1–2 µm)	DNA codes for the synthesis of proteins in the cytoplasm; pores in the envelope allow large molecules in (e.g. enzymes) and out (e.g. RNA); nucleolus synthesises ribosomal RNA and manufactures ribosomes
Ribosomes	Small bodies (20–25 nm) of protein and rRNA either free in the cytoplasm or attached to rough endoplasmic reticulum	Site of protein synthesis
Endoplasmic reticulum (ER)	Membrane system of sacs and tubes permeates the cytoplasm; flattened rough ER is studded with ribosomes; tubular smooth ER lacks ribosomes	Proteins made in the ribosomes accumulate in the rough ER and are passed onto the Golgi apparatus; smooth ER is involved with lipid metabolism

Organelle	Structure	Function
Golgi apparatus	A stack of membrane-bound sacs (cisternae); forming face has vesicles from the rough ER joining; mature face has vesicles pinching off	Dynamic structure in which proteins are modified (may have carbohydrate attached to form glycoproteins) and packaged into vesicles either for secretion by exocytosis or for delivery elsewhere in the cell
Lysosomes	Vesicles produced by the Golgi apparatus which contain hydrolytic enzymes	Lysosomes combine with membrane-bound degenerate organelles or ingested particles (e.g. bacteria) to form secondary lysosomes; hydrolytic enzymes digest the contents (see Figure 34)
Mitochondria (singular: mitochondrion)	Sausage-shaped (1 μm wide and up to 10 μm long); surrounded by an envelope, the inner membrane of which is folded to form cristae; fluid-filled matrix; several to thousands per cell	Synthesis of ATP by aerobic respiration
Chloroplasts	Ovoid (2–10 μm in diameter); surrounded by an envelope; elaborate internal membrane system of lamellae with thylakoids stacked into grana; contain lipid droplets and starch grains; occur in some plant cells	Site of photosynthesis; chlorophyll molecules are attached to the lamellae
Vesicles and vacuoles	Bound by a single membrane; vesicles are much smaller than vacuoles; vacuoles are permanent in plant and fungal cells; membrane of the sap vacuole in plant cells is called the tonoplast	Vesicles may be used for storage and transport of substances (e.g. transport to and from the cell-surface membrane or within the cytoplasm); vacuoles are for storage of water and ions
Microtubules	Tubular (25 nm in diameter); formed from the protein tubulin; occur within centrioles (as nine triplets of microtubules in a circular arrangement) and throughout the cytoplasm; animal and fungal cells contain a pair of centrioles	Centrioles form the spindle fibres during cell division of animal and fungal cells; microtubules also form part of the cytoskeleton and allow movement of cell organelles
Plasmodesmata (singular: plasmodesma)	Strands of cytoplasm between neighbouring plant cells that pass through pores in the walls	Facilitate transport of materials between adjacent cells in plants

Tip Students often confuse secretory vesicles and lysosomes. The Golgi apparatus produces *both*. However, their roles are quite distinct. Secretory vesicles are moved to the cell surface membrane and their contents are exocytosed (see p. 55). Lysosomes remain in the cell where they are involved in intracellular digestion (see Figure 34).

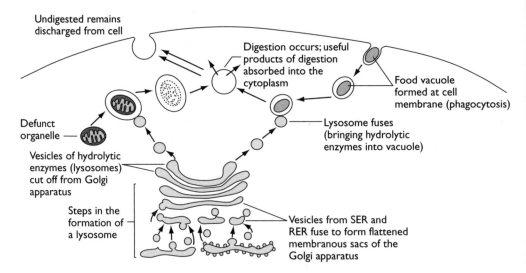

Figure 34 The role of lysosomes

The prokaryotic cell

Prokaryotic cells are particularly small (1–10 µm). They lack a nucleus and other organelles bound by membranes (see Figure 35).

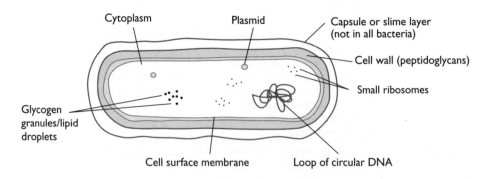

Figure 35 A generalised prokaryotic cell

Prokaryotic and eukaryotic cells are compared in Table 12.

Table 12 Comparison of prokaryotic and eukaryotic cells

Prokaryotic cell	Eukaryotic cell
Small cells — 1–10 µm	Large cells — 10–100 µm
No membrane-bound organelles	Nucleus, mitochondria, endoplasmic reticulum, Golgi apparatus and chloroplasts (in plants) present
Small ribosomes — 20 nm in diameter (70 S)	Large ribosomes — 25 nm in diameter (80 S)
Single circular DNA molecule, without associated protein; the region in the cytoplasm containing the DNA is called the nucleoid	DNA as several linear molecules associated with protein (histones) to form chromosomes; these are contained within a membrane-bound nucleus
Plasmids (small circular pieces of DNA outside the main DNA molecule) usually present	No plasmids
Peptidoglycan cell wall	Cellulose cell wall present in plant cells; and chitin cell walls in fungal cells
No microtubules absent	Microtubules present and organised into centrioles in animal cells
Slimy outer capsule may be present	No capsule

Viruses

Viruses lack cytoplasm and are not cells. They consist of a nucleic acid core surrounded by a protein coat. The nucleic acid ultimately acts as a coding device for the production of new viral particles. There are a number of types (see Figure 36).

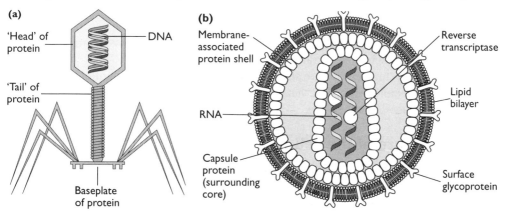

Figure 36 (a) A bacteriophage; (b) The human immunodeficiency virus

Bacteriophages

Bacteriophages (phages) consist of a core of DNA bounded by a protein coat. Phages invade bacteria and the phage DNA codes for the production of new viral proteins (to make new coats). The phage DNA replicates to form many copies which are then packaged in the new protein coats. Eventually the bacterial cells burst to release many new phages.

Human immunodeficiency virus (HIV)

HIV consists of a core of RNA bounded by a protein coat and a lipid bilayer containing glycoproteins. It belongs to a group of viruses containing RNA that are known as **retroviruses**. They contain the enzyme **reverse transcriptase** which catalyses the synthesis of viral DNA from the RNA. Viral DNA ensures that viral protein and new copies of RNA are made. HIV invades a type of lymphocyte (helper T-cell) and so may weaken the immune system, thereby causing AIDS — acquired immune deficiency syndrome.

Practical work

Examine photomicrographs and electron micrographs (TEM/SEM):
- recognise cell structures from photomicrographs and electron micrographs (TEM/SEM)
- calculate true size (in μm) and magnification, including the use of scale bars

Membrane structure and function

The membranes within cells and surrounding them (the cell-surface or plasma membrane) consist of a phospholipid bilayer with associated proteins. In an aqueous environment, phospholipids arrange themselves spontaneously into bilayers so that the hydrophobic 'tail' regions are shielded from the surrounding polar fluid. In cells, this causes the more hydrophilic 'head' regions to associate with the cytoplasmic and extracellular faces on either side.

The structure of the cell-surface membrane

The **phospholipid bilayer** is the basic structure of the cell-surface membrane. There are also **cholesterol** molecules in among the hydrocarbon tails. **Proteins** are attached to the bilayer (**extrinsic**), embedded into one layer (**intrinsic**) or span both layers (**intrinsic** and **transmembranal**). The phospholipids in the cell membrane are constantly moving while the proteins are scattered among them, so that the structure proposed is called the **fluid-mosaic model**.

The membrane also contains **polysaccharides** bound either to the proteins (**glycoproteins**) or to lipids (**glycolipids**). The polysaccharides are on the outer face only, where they form a fringe called the **glycocalyx**.

The fluid-mosaic structure of membranes is illustrated in Figure 37.

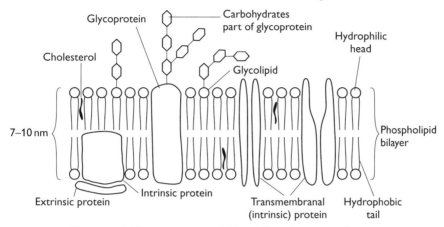

Figure 37 The structure of the cell-surface membrane

Membrane fluidity

A number of factors influence the fluidity of the membrane:

- The more phospholipids with **unsaturated hydrocarbon chains** there are, the more fluid is the membrane. The 'kinks' in the unsaturated hydrocarbon tails prevent them from packing close together so more movement is possible.
- Phospholipids with **longer hydrocarbon chains** will decrease the fluidity of the membrane, since attractive forces among the tails will be greater.
- Fluidity is influenced by **temperature**. The membrance is more fluid at high temperature and less fluid at low temperature as the phospholipid bilayer 'freezes' into a gel (or solid-like) state.
- **Cholesterol** acts as a temperature stability buffer. At high temperature, cholesterol provides additional binding forces and so decreases membrane fluidity. At low temperature, cholesterol keeps the membrane in a fluid state by preventing the phospholipids from packing too close together and 'freezing'.

Cell recognition and cell receptors

Glycoproteins and glycolipids have important roles in **cell-to-cell recognition** and as **receptors** for chemical signals. The glycocalyx allows cells to recognise each other and, therefore, group together to form tissues. Glycoprotein receptors and signalling molecules (e.g. hormones) fit together because they have complementary shapes.

Membrane enzymes

Many of the proteins in the membrane are enzymes. The membrane provides the attached enzymes with improved stability.

Movement of substances in and out of the cell

The cell-surface membrane acts as a barrier between the cytoplasm and the extra-cellular fluid, though exchange of substances takes place across it. The route taken to cross the membrane and the mode of transport depend on a number of factors:

- The **size** of the molecule — very small molecules can slip between the phospho-lipid molecules, large particles can only move in or out by cytosis (bulk transport)
- The **polarity** or **non-polarity** of the substance — non-polar (and lipid-soluble) molecules move through the phospholipid bilayer; polar substances move through the transmembranal proteins
- The **concentration** of the substance either side of the membrane — substances move from high to low concentration by diffusion; if movement against the concentration is required, then active transport is needed.

Passive transport across the cell-surface membrane

Passive movement through the membrane occurs down a concentration gradient and does not require energy expenditure.

Diffusion

Non-polar molecules, such as lipid-soluble vitamins (e.g. vitamins A and D), steroid hormones, the respiratory gases oxygen and carbon dioxide, and very small polar molecules such as water and urea, move through the phospholipid bilayer down their concentration gradients. Ions, which are polar and cannot move through the phospholipid bilayer, and water may diffuse freely through the hydrophilic core of **channel proteins**.

Facilitated diffusion

Carrier proteins may selectively transport ions and molecules with charged groups, such as glucose and amino acids. The substance binds to a site on the protein, which changes shape to bring the substance through the membrane. Movement occurs down the concentration gradient, but being assisted by a carrier is called facilitated diffusion.

Osmosis

Osmosis is the diffusion of water across a differentially permeable membrane (water moves through more easily than solutes). Water diffuses through the phospho-lipid bilayer and through special channel proteins, known as aquaporins. Water moves from an area of higher **water potential** (ψ) to an area of lower water potential. Pure water (at standard temperature and pressure) is defined as having a water potential of zero. The addition of **solutes** to water lowers its potential (makes it more negative), just as an increase in **pressure** increases its potential (makes it more positive).

water potential of a cell (ψ_{cell}) = solute potential (ψ_s) + pressure potential (ψ_p)

The water potential is a measure of the free energy of the water molecules in a system. The water potential of pure water is zero because all the molecules are 'free'. In solutions, some of the water molecules form shells around the solutes and are, therefore, no longer free (see Figure 38). This lowers the water potential — it becomes negative. Cytoplasm contains solutes, so it has a negative water potential. Therefore, when cells are placed in water, water moves into the cells by osmosis.

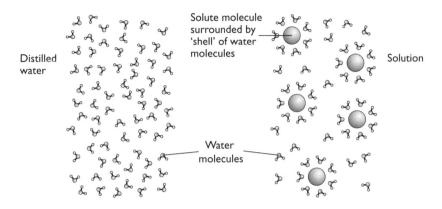

Figure 38 How solutes reduce the water potential of a solution

When placed in dilute (hypotonic) solutions, animal cells, such as red blood cells, take up water and swell until they burst (**lyse**). However, the cells of prokaryotes, fungi and plants have rigid walls that prevent them from bursting. The pressure created by the swelling cell increases to a point when water can no longer enter. A swollen plant cell is said to be **turgid**.

When red blood cells are placed in a concentrated (hypertonic) solution, the cells lose water by osmosis, shrink and become **crenated**. In hypertonic solutions, cells with walls also shrink — the cell wall cannot protect them from water loss by osmosis. As they shrink, the cells lose contact with their cell walls. In plant cells, this is known as **plasmolysis** and the point at which the cytoplasm just begins to lose contact with the cell wall is called **incipient plasmolysis**.

Water potential, the solute potential of the cell contents and the pressure potential of a plant cell all change as the cell takes up or loses water osmotically (see Figure 39).

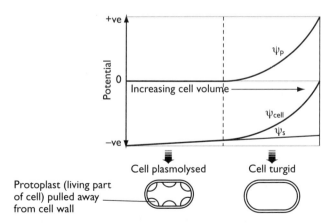

Figure 39 Changes in pressure potential (ψ_p), solute potential (ψ_s) and water potential (ψ_{cell}) of a plant cell as it takes up water osmotically

Active transport across the cell-surface membrane

Active transport causes substances to move across a membrane from a low concentration to a high concentration against the concentration gradient. This requires energy, and the **carrier proteins** involved are referred to as **pumps**. Each carrier protein is specific to just one type of ion or molecule. The substance attaches to a site on the protein (they have complementary shapes) and, with the energy from ATP, the protein changes shape and moves the substance through the membrane.

The movement of substances across the cell-surface membrane is illustrated in Figure 40.

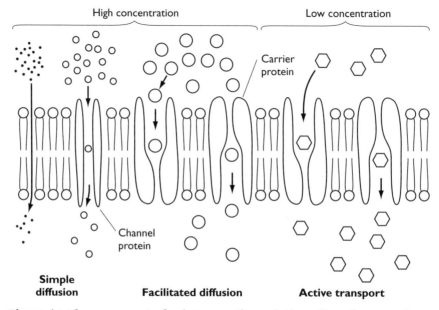

Figure 40 The movement of substances through the cell-surface membrane

Cytosis: bulk transport into and out of the cell

Substances can move into and out of a cell without having to pass through the cell-surface membrane. This involves the bulk transport into the cell (**endocytosis**) or out of the cell (**exocytosis**) of substances too large to be transported by protein carriers.

Endocytosis

During endocytosis the cell-surface membrane invaginates and the membrane folds round the substance to form a vacuole or vesicle that enters the cytoplasm while the cell-surface membrane reforms. There are two main types:

- **Phagocytosis** is the uptake of solid particles into the cell within vacuoles. Examples include the ingestion of bacteria by polymorphs (a type of white blood cell) and the removal of old red blood cells from circulation by the Kupffer cells of the liver.

- **Pinocytosis** is the uptake of solutes and large molecules (such as proteins) into the cell within vesicles.

Exocytosis

During exocytosis, **secretory vesicles** move towards and fuse with the cell-surface membrane releasing their protein contents out of the cell. Exocytosis is also involved in removal of the waste products of digestion from cells.

Cytosis is illustrated in Figure 41.

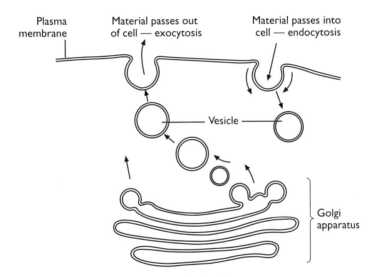

Figure 41 Cytosis: bulk transport

Practical work

Measure the average water potential of cells in a plant tissue:
- use a weighing method for potato or other suitable tissue
- calculate the percentage change in mass
- determine the average water potential from a graph of percentage change in mass against solute potential of immersing solution

Measure the average solute potential of cells at incipient plasmolysis:
- use of onion epidermis or other appropriate material
- calculate percentage plasmolysis
- determine the average solute potential from a graph of percentage plasmolysis against solute potential of the immersing solution; at 50% plasmolysis, the average pressure potential is zero

The cell cycle, mitosis and meiosis

The formation of new cells, for example in the development of a multicellular organism, involves the production of additional cell contents before a cell can divide. The pattern of events is called the **cell cycle**.

The cell cycle

Actively dividing eukaryotic cells pass through a series of stages known collectively as the cell cycle:

- **interphase** — two growth or **gap phases (G1** and **G2**) separated by a **synthesis (S) phase**
- **mitosis** — a nuclear division during which the chromosomal material is partitioned into daughter nuclei
- **cytokinesis** — the cell divides into two daughter cells

Interphase

This is an intense period of metabolic activity as the cell synthesises new components such as organelles and membranes, and new proteins and DNA. The production of new cellular proteins occurs throughout interphase; DNA synthesis occurs during the S phase. Histones, proteins that bind to and support the DNA within the chromatids, are also produced during the S phase. The DNA and chromatids formed are identical and remain attached until separated during mitosis. The S phase separates the G1 and G2 phases. Among the proteins synthesised in G2 is tubulin, which combines to form the microtubules of the spindle.

Mitosis

During mitosis, different stages are recognised (see Figure 42).

Prophase

- The chromatin condenses to form the chromosomes
- The centrioles (in animal cells) move towards opposite poles
- The **spindle** begins to form
- As each chromosome continues to condense, two chromatids, joined at the **centromere**, become apparent

Prophase (late)

Metaphase

- The nuclear envelope breaks down
- Spindle formation is completed as microtubules extend, forming the fibres
- The microtubules of the spindle attach to the centromere of each chromosome
- The chromosomes (chromatid pairs) are moved by the microtubules onto the equator of the spindle

Metaphase

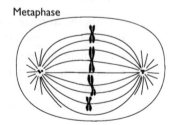

Anaphase

- The centromeres divide
- The spindle fibres pull the centromeres of sister chromatids apart
- The sister chromatids move towards opposite poles

Telophase

- Each chromatid is now a separate chromosome
- The two groups of chromosomes reach opposite poles of the cell
- A new nuclear envelope forms around each group

Anaphase

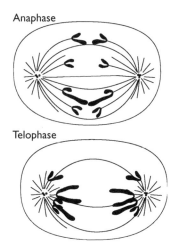

Telophase

Figure 42 The stages of mitosis

Cytokinesis

At the end of mitosis, the cytoplasm is separated and the cell divides during cytokinesis to form two daughter cells. The process differs in animal and plant cells. In the animal cell, a **cleavage furrow** forms as protein microfilaments pull the cell surface membrane in along the equator; the furrow deepens and when the membranes fuse the cell is cleaved into two. In plant cells, the cell wall prevents cleavage. In the plant cell, the Golgi bodies (known as dictyosomes) produce vesicles that collect and fuse together to form an equatorial **cell plate**. The vesicles secrete the material of the middle lamella on each side of which a new cellulose cell wall is laid down.

Genes, chromosomes and ploidy

Specific lengths of DNA represent the **genes** (codes for the synthesis of proteins). The positions of genes on a chromosome are called **genetic loci** (singular: locus). During interphase, much of the DNA is unwound (**euchromatin**) in order to allow easy access to the code; some DNA remains condensed (the **heterochromatin**) because access to it is not required. At the onset of nuclear division (whether mitosis or meiosis) all the chromatin becomes condensed (supercoiled); this increases its strength. The physical strength of the chromatids and chromosomes that form is important in preventing DNA breakage when they are pulled apart on the spindle apparatus.

In most plants and animals, the cells of the body each contain two sets of chromosomes, which exist in **homologous pairs**. Each member of a pair is similar in size and shape to the other. More importantly, they have the same genetic loci — they possess **alleles** of the same genes (one from each parent). If the alleles on the homologous chromosomes are the same then the individual is **homozygous** for that particular characteristic; if they are different then the individual is **heterozygous**. Cells containing homologous pairs of chromosomes are said to be **diploid** (represented by **2*n***).

During the S phase, DNA replicates and each new DNA molecule associates with protein (histones) to form sister chromatids. Since DNA replication produces identical copies, the chromatids are genetically identical. Homologous chromosomes are genetically different, since at least some of the hundreds of genetic loci will possess different alleles.

The relationship between genes, alleles, chromatids and homologous chromosomes is shown in Figure 43.

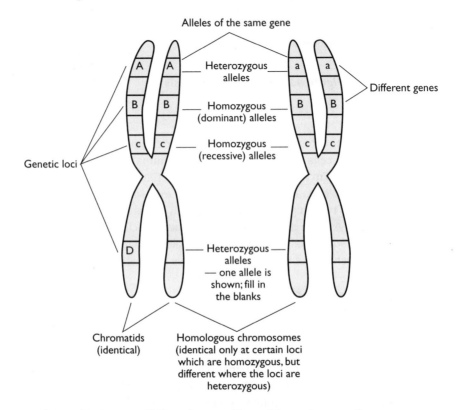

Figure 43 Genes, alleles, chromatids and homologous chromosomes

A cell that contains only one of each type of chromosome is **haploid** (represented by **n**). Some simple organisms (e.g. mosses) contain cells with haploid nuclei but in higher forms it is only the gametes that are haploid.

Meiosis

Meiosis occurs only in diploid cells. It involves the separation of homologous chromosomes during a first meiotic division (**meiosis I**) and the separation of chromatids during a second meiotic division (**meiosis II**). The apparatus for these divisions is the same as in mitosis, so the emphasis in Figure 44 is on points specific to meiosis.

Prophase I

- As chromosomes condense it becomes apparent that homologous chromosomes have paired and lie alongside each other; each pair is known as a **bivalent**
- The chromatids appear; the chromatids in a bivalent are entwined at points called chiasmata (singular: chiasma)
- The chromatids may break at chiasmata and rejoin with a different chromatid, resulting in crossing over or recombination

Metaphase I

- The bivalents move to the equator of the spindle
- Each chromosome of the pair becomes attached to a spindle fibre by its centromere

Anaphase I

- Pulling by the spindle fibres causes the whole chromosomes to move apart towards opposite poles
- The homologous chromosomes are separated; each chromosome still consists of two chromatids

Telophase I

- Chromosomes reach opposite poles of the cell.
- A nuclear membrane forms around each separate group of chromosomes; each nucleus contains the haploid number of chromosomes

Figure 44 The stages of meiosis I

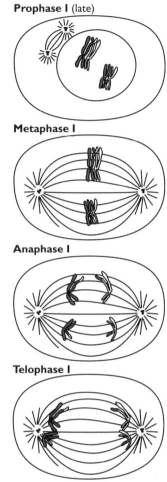

Prophase I (late)

Metaphase I

Anaphase I

Telophase I

Cytokinesis after meiosis I produces two daughter cells. Within each, meiosis II follows:

- New spindles begin to form at right angles to the old spindle (**prophase II**).
- Chromosomes consisting of pairs of chromatids (now different because of crossing over) are arranged along the equator (**metaphase II**).
- Sister chromatids are split at the centromere and pulled to opposite poles (**anaphase II**).
- Each group of separated chromosomes becomes enclosed within a nuclear envelope (**telophase II**).

Cytokinesis follows. The overall result of meiosis is the production of four haploid daughter cells, each of which is genetically different from the others.

The significance of mitosis

Mitosis produces genetic constancy:

- The daughter cells possess the *same* chromosome number as each other and as the parent cell. Mitosis can occur in either diploid cells (e.g. during the development of a mammal) or haploid cells (e.g. in the growth of mosses).

- The daughter cells are *genetically identical.* Mitosis has a key role in asexual reproduction, producing genetically identical individuals (**clones**).

The significance of meiosis

Meiosis produces change:
- Meiosis is the type of nuclear division that transforms the diploid condition to the haploid condition. This is vital in life cycles where **fertilisation** involves the fusion of gametes (haploid cells) to form the zygote (a diploid cell).
- Meiosis produces daughter cells which are *genetically different.* This genetic variation occurs as a result of **crossing over** of chromatid pieces (during prophase I) and of the **independent assortment of bivalents** (during metaphase I).

Crossing over

Crossing over (see Figure 45) occurs as a result of chiasmata formation between the chromatids of the homologous pairs during late prophase I. A piece of chromatid from one chromosome swaps places with a piece of chromatid of the homologous partner. It results in each chromosome having a different combination of alleles (called **recombinants**) from that which occurred originally.

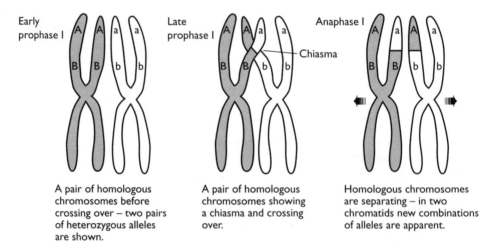

A pair of homologous chromosomes before crossing over – two pairs of heterozygous alleles are shown.

A pair of homologous chromosomes showing a chiasma and crossing over.

Homologous chromosomes are separating – in two chromatids new combinations of alleles are apparent.

Figure 45 Crossing over and genetic variation

Independent assortment

During metaphase I, bivalents are arranged at *random* on the equator of the spindle. This means that the orientation of any one homologous pair is not dependent on the orientation of any other pair. When the homologous chromosomes are pulled apart at anaphase I, a chromosome of one pair is equally likely to be separated along with either member of any other homologous pair.

Independent assortment is illustrated in Figure 46.

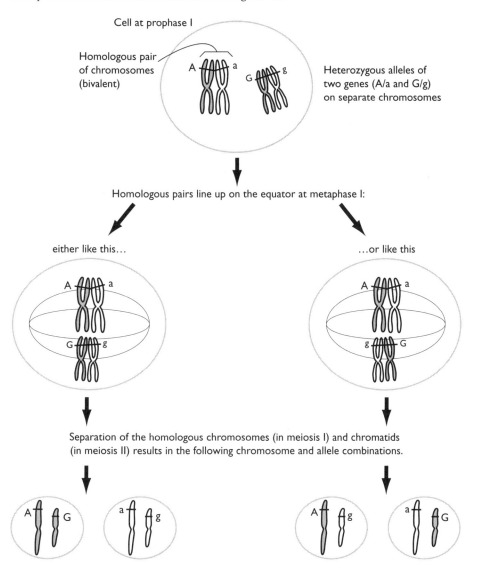

Figure 46 Independent assortment and genetic variation

Mitosis and meiosis are compared in Table 13.

Table 13 Comparison between mitosis and meiosis

Mitosis	Meiosis
One division, producing two daughter cells	Two divisions, producing four daughter cells
Parent cell may be either diploid or haploid; daughter cells have the same chromosome number as the parent cell	Parent cell is always diploid; daughter cells are haploid
Homologous chromosomes (if the parent cell is diploid) do not associate during prophase	Homologous chromosomes pair, forming bivalents during prophase I
No chiasmata formation	Chiasmata form between the chromatids of the homologous chromosomes during prophase I
Daughter cells are genetically identical	Daughter cells are genetically different

Practical work
Prepare and stain root tip squashes and examine prepared slides or photographs of the process of mitosis:
- recognise chromosomes at different stages of cell division
- identify the stages of mitosis
- examine prepared slides or photographs of the process of meiosis
- identify the stages of meiosis

Tissues and organs

Animals and plants are multicellular — they are made up of large numbers of cells. Cells become specialised according to their function. **Tissues** are made up of many cells that perform one or several functions. Often the cells are of the same type. For example, epithelia are sheets of cells that line organs and separate internal tissues from air, blood, food and waste travelling through tubes in the body. In plants, the epidermis secretes a waxy cuticle to protect the plant from desiccation. **Organs** are structures made of several tissues that work together to carry out a number of functions. The **leaf** contains epidermis for protection, mesophyll for photosynthesis and gaseous exchange, xylem for transport of water and phloem for transport of sucrose. The **ileum** is the organ, in the small intestine, that is concerned with the final stages of digestion, the absorption of the products of digestion into blood vessels for transport to other parts of the body, and the movement of undigested material along to the large intestine. Many body processes are performed by groups of organs working together — **organ systems**. For example, in the digestive system of a mammal, the mouth, oesophagus, stomach, small and large intestines, liver, gall bladder and pancreas work together to digest and absorb food and eliminate undigested material.

The ileum

The ileum is the region of the small intestine where digestion is completed and where most absorption of the products of digestion occurs. Numerous folds in the wall of the ileum increase its surface area. The folds themselves have numerous **villi** (singular: villus) and the lining cells possess **microvilli**, all of which means that there is a vast surface area for digestion and absorption.

Structurally the ileum consists of tissues in distinct layers: **mucosa**; **muscularis mucosa**; **submucosa**; **muscularis externa** and **serosa** (outermost) (see Figure 47).

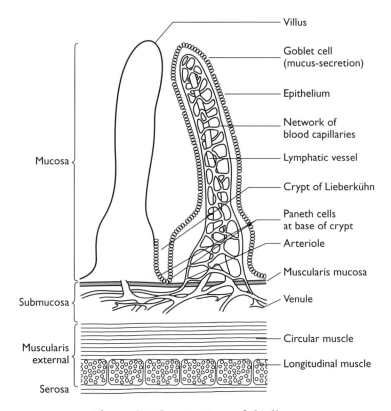

Figure 47 The structure of the ileum

The functions of the tissues are shown in Table 14.

Table 14 The functions of tissues in the ileum

Tissue	Function
Columnar epithelium (within the mucosa)	This layer has column shaped cells and lines the intestine. On their free surfaces, the cells have microvilli, forming a brush border. Since digestive enzymes are bound to the membrane of the microvilli, this provides a huge surface area for digestion and for the absorption of the products of digestion. Some substances are taken up partly by diffusion and partly by active transport; others are taken up by pinocytosis. There are numerous mitochondria to aid active transport. The cells of the epithelium are short-lived (see crypts of Lieberkühn).
Goblet cells (within the epithelium)	These cells secrete mucus. Mucus is slimy. It protects the epithelium from the action of digestive enzymes and lubricates the lining as solid material is pushed along.
Villi (within the mucosa)	These finger-like projections increase the surface area for the absorption of the products of digestion. The villi contain blood capillaries into which amino acids and monosaccharides are absorbed, and lacteals (blind-ending lymph vessels) into which fats are absorbed.
Crypts of Lieberkühn (within the mucosa)	These intestinal glands are found at the bases of the villi. The cells along the sides secrete mucus. The cells (stem cells) lining the bottom of the crypts are in a state of continuous division; new cells are continuously being pushed up by the division of cells deeper down. After a life of several days within the epithelium, the cells are pushed to the tips of the villi where they are sloughed off. Paneth cells are also present at the base of the crypts. Their function is to defend the actively dividing cells against microbes in the small intestine.
Muscularis mucosa	The muscle fibres contract to cause movement of the villi, so improving contact with the products of digestion.
Submucosa	The submucosa contains blood vessels including venules of the hepatic portal vein (carrying blood to the liver) and lymphatic vessels, supported by connective tissue.
Muscularis externa	The muscularis externa consists of circular muscle (innermost) and longitudinal muscle. Contractions of longitudinal muscle causes pendular movement of the gut while contraction of circular muscle may result in local constrictions, both of which churn the food. Coordinated contractions of the circular muscle push food along the gut by peristalsis.
Serosa	This outer layer of connective tissue serves to protect and support the gut.

The leaf

The leaf has a large surface area which maximises the absorption of light for photosynthesis. It is thin, and so photosynthesising cells are not far from the leaf surfaces where light absorption and gaseous exchange occur.

Structurally, the leaf consists of epidermal layers either side of a middle layer of mesophyll and vascular tissues (see Figure 48).

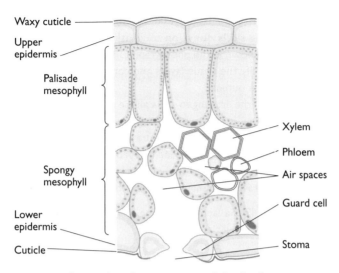

Figure 48 The structure of the leaf

The functions of the tissues are shown in Table 15.

Table 15 The functions of tissues in the leaf

Tissue	Function
Upper epidermis	The cells of the upper epidermis lack chloroplasts since their role is protective. They secrete a waxy cuticle that provides waterproofing and reduces water loss.
Palisade mesophyll	The palisade layer, in the upper half of the leaf, has layers of tightly packed cells, each with abundant chloroplasts. It is adapted for maximal light absorption. This is the main photosynthetic region of the leaf.
Spongy mesophyll	The mesophyll in the lower half of the leaf contains large air spaces. Gaseous exchange between these air spaces and the atmosphere can take place via numerous pores (stomata). Spongy mesophyll cells also contain chloroplasts and are photosynthetic.
Xylem vessels (within vascular bundles)	Xylem vessels supply the leaf with water and inorganic ions.
Phloem sieve tubes (within vascular bundles)	Phloem sieve tubes transport sucrose away from the leaf.
Lower epidermis	The cells lack chloroplasts. The cuticle secreted on the lower surface is thinner than that on the upper surface since it is not exposed directly to the sun.
Stomata	The lower epidermis contains numerous stomata which allow gaseous exchange. They also allow water vapour to diffuse easily out of the leaf. Each stoma (singular of stomata) is surrounded by a pair of guard cells which cause it to close at night and so water loss by transpiration is minimised.

Tip In an AS paper, you may be asked to make a labelled drawing from a photograph. A **drawing skills question** could ask you to 'draw a block diagram to show the tissue layers shown in the photograph'. You will be tested on:

- your ability to identify the tissue layers and construct a block drawing whereby only tissue layers are outlined
- the completeness of the drawing to show all the tissues obvious in the photograph
- how true the drawing is to the photograph provided (and not just a textbook diagram of the feature) and that the drawing is the same magnitude as the photograph (or that a scale is provided if appropriate)
- the position and proportionality of the tissue layers
- the quality of the drawing so that clear, smooth and continuous lines are drawn

As an example, Figure 49 is a labelled drawing of a photograph of the midrib region of a privet leaf that shows the tissue layers.

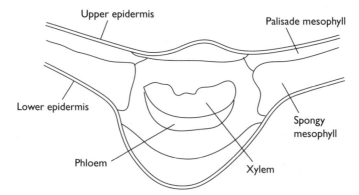

Figure 49 A block diagram of the midrib region of a privet leaf

Practical work

Examine stained sections of the ileum using the light microscope or photographs of the same:

- recognise the villi (and associated blood capillaries and lacteals), crypts of Lieberkühn (and Paneth cells), mucosa, columnar epithelium, goblet cells, muscularis mucosa, submucosa, muscularis externa, and serosa

Examine sections of a mesophytic leaf using the light microscope or photographs of the same:

- recognise the epidermal layers, waxy cuticles, palisade mesophyll, chloroplasts, spongy mesophyll, vascular vessels with xylem and phloem, and guard cells and stomata

Make accurate drawings of sections of the ileum and the leaf to show the tissue layers:

- draw block diagrams of tissues within the ileum and the leaf

Questions
&
Answers

This section consists of two exemplar papers constructed in the same way as your AS Unit 1 examination papers. There are questions that assess straightforward knowledge and understanding, some of which require you to apply your understanding to novel situations and a few that assess your knowledge of practical techniques. There is a variety of question styles. Each paper has a total of 75 marks and you have 1 hour 30 minutes to attempt all the questions.

Following each question, there are answers provided by two students — Candidate A and Candidate B. These are real responses. Candidate A has made mistakes that are often encountered by examiners and the overall performance might be expected to achieve a grade C or D. Candidate B has made fewer mistakes. The overall performance is good and is of grade A or B standard.

Examiner's comments

These are preceded by the icon 𝒆. They provide the correct answers and indicate where difficulties for the candidate occurred. Difficulties may include lack of detail, lack of clarity, misconceptions, irrelevance, poor reading of questions and mistaken meanings of examination terms. The comments suggest areas for improvement.

Using this section

You could simply read this section, but it is always better to be *active* in developing your examination technique. One way to achieve this would be to:
- try all the questions in Exemplar Paper 1 before looking at candidates' responses or the examiner's comments, allowing yourself 1 hour 30 minutes —remember to follow the suggestions in the introduction
- check your answers against the candidates' responses and the examiner's comments
- use the answers provided in the examiner's comments to mark your paper
- use the candidates' responses and the examiner's comments to check where your own performance might be improved

You should then repeat this for Exemplar Paper 2.

Section A

Question 1

Identify *four* distinct differences between mitosis and meiosis.   (4 marks)

Total: 4 marks

Candidates' answers to Question 1

Candidate A

Mitosis consists of a single division, while there are two cell divisions in meiosis ✓.

Mitosis produces two cells, while meiosis produces four [equivalent point awarded above]

In mitosis, the daughter cells are identical, while in meiosis the cells are genetically different ✓.

The daughter cells in mitosis are diploid, while the daughter cells in meiosis are haploid ✗.

> 🖉 The first two answers, regarding number of divisions and the number of daughter cells produced are not sufficiently distinct, so they only gain 1 mark. The third answer is correct. The fourth answer is incorrect. Meiosis always produces haploid cells, but mitosis produces diploid or haploid cells depending on the ploidy of the parent cell — it maintains the constancy of chromosome number. The candidate scores 2 out of 4 marks.

Candidate B

Homologous chromosomes do not pair in mitosis, but do in meiosis ✓.

Chiasmata are not formed in mitosis, but are formed in meiosis ✓.

In mitosis, single chromosomes line up on the equator at metaphase; in meiosis, homologous pairs of chromosomes assemble on the equator at metaphase I ✓.

Mitosis produces genetically identical cells, while meiosis produces daughter cells that are genetically different ✓.

> 🖉 All the answers are correct, scoring the maximum 4 marks. Note that these are not the only possible correct answers.

> 🖉 **Overall, Candidate A scores 2 marks and Candidate B scores 4.**

Question 2

The molecular diagram below illustrates a reversible reaction involving two amino acids.

(a) (i) **Name the type of reactions labelled X and Y.** (2 marks)

(ii) **Name the type of bond labelled Z.** (1 mark)

(iii) **Name the final product of reaction X.** (1 mark)

(b) **Describe what is meant by the primary structure of a protein.** (1 mark)

Total: 5 marks

Candidates' answers to Question 2

Candidate A

(a) (i) X — condensation ✓
    Y — hydration ✗

   The candidate has confused the word hydration with hydrolysis. Hydration is the addition of water but this is breakdown by chemical reaction with water. The candidate scores 1 mark.

(ii) Peptide bond ✓

   This is correct, for 1 mark.

(iii) A dipeptide ✓

   This is correct, for 1 mark.

(b) A chain of amino acids ✗

🖉 This is not a sufficiently precise answer to earn a mark. The order or sequence of the amino acids in the chain would be better wording.

Candidate B

(a) (i) X – condensation ✓
   Y – hydrolysis ✓

🖉 Both answers are correct. The candidate scores 2 marks.

(ii) Peptide bond ✓

🖉 This is correct, for 1 mark.

(iii) A dipeptide ✓

🖉 This is correct, for 1 mark.

(b) The sequence of bases ✗ and therefore the amino acid sequence.

🖉 The sequence of amino acids is correct but not the sequence of bases leading to it. This is the genetic code — the order of base triplets that determines the primary structure of a protein. No mark can be awarded.

🖉 **Overall, Candidate A scores 3 marks and Candidate B scores 4.**

Question 3

The photograph below is an electron micrograph of part of a eukaryotic cell.

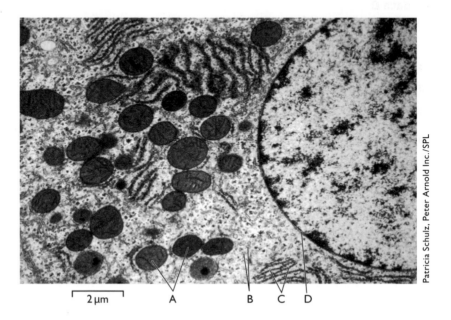

Patricia Schulz, Peter Arnold Inc./SPL

2 μm   A   B   C   D

(a) Identify the features labelled A to D. (4 marks)

**(b) The photograph has a scale bar indicating 2 μm. Use this to calculate
the magnification of this electron micrograph. Show your calculations.** (3 marks)

Total: 7 marks

Candidates' answers to Question 3

Candidate A

(a) A — lysosomes ✗
    B — vesicles ✓
    C — rough ER ✓
    D — nuclear membranes ✓

    ✒ The answer to A is incorrect. The internal cristae are obvious in the photograph
    and so A are mitochondria. The other answers are correct, scoring 3 marks.

(b) size of image = real size × magnification

    ✒ The candidate has remembered the equation that describes the relationship
    between the dimensions of an object and its magnification, but has not been able
    to apply it to the magnification of the electron micrograph. The candidate fails
    to score.

Candidate B

(a) A — mitochondria ✓
  B — ribosomes ✗
  C — rough endoplasmic reticulum ✓
  D — nuclear membranes ✓

📝 The answer to B is incorrect. Ribosomes are solid structures and also smaller, as is apparent on the rough endoplasmic reticulum where they are clearly membrane-bound. The other answers are correct, so the candidate scores 3 marks.

(b) The scale bar is 14 mm long ✓ which is 14 000 μm ✓, so magnification is 14 000 ÷ 2 = 7000 times ✓

📝 All stages in the calculation (measurement of scale bar, unit conversion and determination of magnification) are correct and clearly shown. Showing each stage is important because if a 'slip' is made at any stage, marks can still be awarded for the correct procedure. The candidate gains all 3 marks.

📝 **Overall, Candidate A scores 3 marks and Candidate B scores 6.**

Question 4

The enzyme lactase catalyses the hydrolysis of lactose into galactose and glucose.

(a) Explain why lactase acts only on lactose and not on other disaccharides.

(1 mark)

The graph below shows the effect of varying pH on lactase activity when in solution (o) and when immobilised (●).

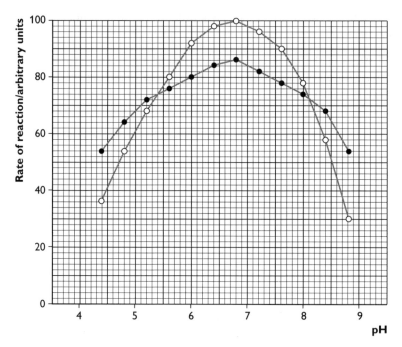

(b) Both in solution and when immobilised, lactase activity was maximal at a particular pH.

(i) Using the information in the graph, determine the optimal pH for lactase. (1 mark)

(ii) Explain why enzymes exhibit an optimal pH. (1 mark)

(c) (i) Describe two differences in the activity of lactase in solution and when immobilised, as shown in the graph. (2 marks)

(ii) Explain the differences you described in your answer to (c) (i). (2 marks)

Total: 7 marks

Candidates' answers to Question 4

Candidate A

(a) The enzyme lactase has an active site that is complementary only to the shape of lactose ✓.

  🖉 This is correct, for 1 mark.

(b) (i) pH 6.8 ✓

  🖉 This is correct, for 1 mark.

  (ii) It is the point at which there are most successful collisions ✗.

  🖉 This is simply irrelevant. Temperature and concentration are factors that can affect the rate of collisions between enzyme and substrate molecules, but not pH. No mark can be awarded.

(c) (i) The enzyme in solution has a higher maximum rate of reaction than the immobilised enzyme ✓.

  The immobilised enzyme starts off and finishes the experiment with a higher rate of reaction ✗.

  🖉 The first answer is correct and scores 1 mark. The second is incorrect because it assumes a time element, which is not evident in the experiment. The independent variable is pH, not time.

  (ii) In solution, the substrate bonds freely with the enzyme, but does not when the enzyme is immobilised ✓.

  When immobilised, the enzymes are more stable and are less affected by the pH ✗.

  🖉 The first answer is sufficient for a mark. The second is not, since there is no attempt to *explain* the increase in stability.

Candidate B

(a) Because its active site is specific to lactose molecules only ✗.

  🖉 This does not *explain* the specificity — for example, there is no mention of the lactase active site being complementary in shape to lactose. The candidate fails to score.

(b) (i) pH 6.6 ✗

  🖉 Incorrect. Always try to make time to check through numerical work. This is obviously a 'slip' — nevertheless, no mark can be awarded.

(ii) At the optimum pH the enzyme's active site is at its most complementary shape. At extremes of pH the ionic bonds are affected, altering the shape of the active site ✓.

🖉 This is an excellent answer and scores the mark.

(c) (i) The immobilised enzyme has a lower rate of reaction at the optimum pH ✓.

The immobilised enzyme is more active than the enzyme in solution at the extremes of pH ✓.

🖉 Both answers are correct, for 2 marks.

(ii) Immobilised enzymes are physically bound within a substance (e.g. alginate) so the substrate cannot move as freely to the enzyme ✓.

Immobilised enzymes are more stable because they are entrapped within a support material ✓.

🖉 These are correct well-phrased answers that earn both marks.

🖉 **Overall, Candidate A scores 4 marks and Candidate B scores 5.**

Question 5

An experiment was undertaken to determine the water potential of potato tuber tissue. The procedure was as follows:

(1) Obtain samples of tissue of approximately equal size
(2) Cut tissue samples into slices
(3) Surface-dry slices using filter paper
(4) Weigh tissue sample slices
(5) Add to one of a series of sucrose solutions of known solute potential
(6) Leave for 24 hours, after which surface-dry slices again and reweigh
(7) The change in mass is expressed as a percentage of the initial mass

(a) (i) Explain how equal-sized portions of tissue may be obtained.

(ii) Explain why the tissue sample is cut into slices.

(iii) Explain why the slices of tissue are surface-dried.

(iv) Explain why the change in mass is expressed as a percentage. (4 marks)

The results of the experiment are shown in the graph below.

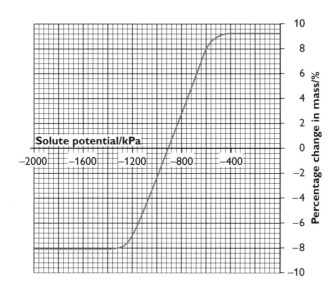

(b) With reference to water potential, explain the change in mass that occurred when the slice of potato tissue was immersed in a sucrose solution of solute potential −600 kPa. (2 marks)

(c) Using the information in the graph, determine the water potential of the potato tissue. Explain your reasoning. (2 marks)

(d) In an evaluation of the experiment the importance of a 'standardised drying technique' was emphasised. Describe what is meant by a standardised drying technique, and explain its importance to the accuracy of the result. (2 marks)

Total: 10 marks

Candidates' answers to Question 5

Candidate A

(a) (i) By using a cork borer ✗

(ii) To increase the surface area for exchange ✓

(iii) To remove any water from the surface that may affect the results ✓

(iv) To make it easier to see the difference ✗

🖉 The use of a cork borer only is not sufficient because the cylinders have to be cut to the same length. The second and third answers, though correct, might have been more fully explained (see Candidate B's responses). The fourth answer is incorrect. Candidate A scores 2 marks.

(b) Water moves from an area of high water potential to an area of low water potential ✓ and so moves out ✗ of the potato.

🖉 Movement of water from high to low water potential is correct, for 1 mark. However, water does *not* move out of the potato tissue, since at −600 kPa there in an increase in the mass of the potato.

(c) −900 kPa ✓. At this point there is no percentage change in mass, which means that the water potential in the tissue and the sucrose solution is equal ✓, so there is no net movement of water.

🖉 Both the determination and the rationale are correct, for 2 marks.

(d) The same pressure is applied to the filter paper when drying the potato discs ✓, so that the results are reliable ✗.

🖉 The description of the standardised drying technique is correct, for 1 mark. However, its use has nothing to do with reliability, which is often confused with accuracy.

Candidate B

(a) (i) Use a cork borer of the same diameter and cut the cylinders to equal lengths ✓

(ii) To provide a greater surface area over which osmosis can take place ✓

(iii) To remove excess surface water that is not within the potato tissue ✓

(iv) To allow changes in mass to be directly compared, since the tissue samples would not be the same mass to start with ✓

🖉 These are excellent well-phrased answers. The candidate scores the full 4 marks.

(b) With a solution of –600 kPa, the potato gains mass and so water must have entered it ✓, which means that the potato tissue had a lower water potential ✓.

🖉 Both points are correct and well explained, for 2 marks.

(c) –850 kPa ✗. The pressure potential is zero, so the water potential is equal to the solute potential ✗.

🖉 The candidate has not accurately read the x-axis scale and needs to check numerical work. The reasoning given is not that for determining the water potential, but for determining the solute potential of, for example, epidermal tissue. The candidate has confused the two experiments and fails to score.

(d) The same pressure has to be applied to the filter paper when drying before both initial and final weighings ✓. If not sufficiently dried before the final weighing, then the weight loss would be less than it should have been ✓.

🖉 These are excellent answers. Candidate B scores both marks.

🖉 **Overall, Candidate A scores 6 marks and Candidate B scores 8.**

Question 6

(a) The active uptake of solutes by cells involves membrane carrier molecules. The carrier combines with the solute and then, utilising the energy of ATP, transfers it to the inner side of the membrane and releases it. The mechanism is represented in the diagram below.

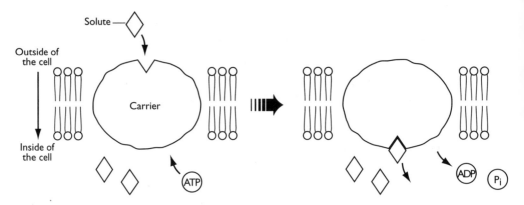

(i) What type of molecule is the carrier? (1 mark)

(ii) For any particular solute that is absorbed actively there is a specific membrane carrier. Explain this specificity. (2 marks)

(iii) Explain how water passes through the membrane. (1 mark)

(b) The uptake of potassium ions by plant tissue placed in potassium chloride solution of different initial concentrations and at two different temperatures was investigated. The results of the experiment are shown in the table below.

Initial concentration of potassium chloride solution/mM	Uptake of potassium ions at 4°C/arbitrary units h^{-1}	Uptake of potassium ions at 18°C/arbitrary units h^{-1}
0	0	0
5	14	30
10	18	38
20	22	48
40	23	50

(i) **Plot the above data, using the most appropriate graphical technique.**

(5 marks)

(ii) **The plant tissue has an internal potassium ion concentration of 50 mM. Explain why the potassium uptake must involve active transport.** (1 mark)

(iii) **Explain the effect of temperature on the rate of potassium ion uptake.** (2 marks)

(iv) **Active transport involves membrane carrier molecules with which the ions combine before being transferred to the inner side of the membrane. Use this information to suggest why at 18°C:**
- **increasing the concentration of potassium ions from 0 mM to 20 mM greatly increases the rate of uptake**
- **increasing the concentration of potassium ions from 20 mM to 40 mM does not appreciably increase the rate of uptake** (2 marks)

(v) **Rubidium ions have similar properties to potassium ions and, when present in the external solution, reduce the rate of potassium ion uptake. Suggest a reason for this observation.** (1 mark)

Total: 15 marks

Candidates' answers to Question 6

Candidate A
(a) (i) Protein ✓

🖉 This is correct, for 1 mark.

(ii) Fat-soluble molecules go through the phospholipid bilayer, and water-soluble molecules go through the hydrophilic channels ✗.

🖉 The candidate has not read the question carefully enough. The question requires an explanation of the specificity of membrane carriers. The candidate fails to score.

(iii) Through the hydrophilic channels ✗.

🖉 This is incorrect — water molecules are small enough to pass through the phospholipid bilayer.

(b) (i)

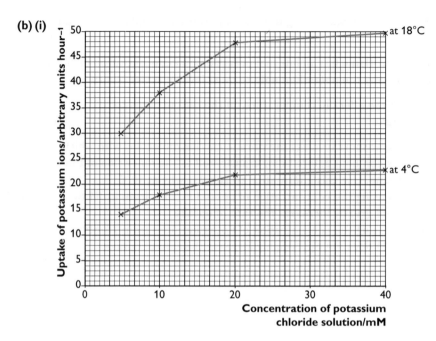

Concentration of potassium chloride solution/mM

🖉 There is no caption to explain the contents of the graph ✗.

The concentration of potassium chloride solution is the independent variable ✓.

Both axes have labels with units of measurement ✓.

The 0,0 points are not plotted and so the initial part of the graph is missing ✗.

The lines for 4°C and 18°C are identified ✓.

The candidate scores 3 of the 5 marks available.

(ii) There is a greater concentration of potassium chloride inside than there is outside, so it can't move by diffusion ✓.

🖉 This is correct, for 1 mark.

(iii) As temperature increases, the rate of ion uptake increases since the ions have more kinetic energy ✗.

🖉 In part (b) (ii) it was established that ion uptake does not occur by diffusion. The answer should relate to the availability of ATP for the operation of the membrane carriers (see Candidate B's response). No marks can be awarded.

(iv) From 0 mM to 20 mM: more potassium ions are available for attachment to the carriers ✓.

From 20 to 40 mM: there are a limited number of carrier proteins to transport the potassium ions ✓.

🖉 The answers to both parts of the graph are correct. The candidate earns 2 marks.

(v) They inhibit the carrier proteins by taking the place of K⁺ ions ✓.

🖉 This is correct, for **1 mark**.

Candidate B

(a) (i) Protein ✓

🖉 This is correct, for **1 mark**.

(ii) Each carrier protein has a receptor site ✓ that has a complementary shape ✓ for the attachment of a specific solute.

🖉 The candidate scores both marks.

(iii) Water molecules are sufficiently small to diffuse across the phospholipid bilayer ✓.

🖉 This is correct, for **1 mark**.

(b) (i)

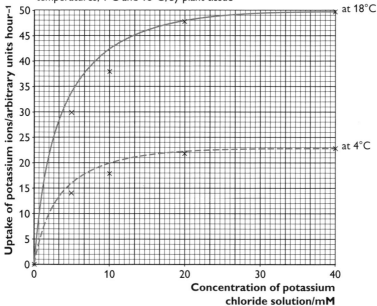

The uptake of potassium ions in different concentrations of KCl solution at two different temperatures, 4°C and 18°C, by plant tissue

🖉 An appropriate caption noting K⁺ uptake, potassium chloride concentration and temperature is included ✓.

The concentration of potassium chloride solution is the independent variable ✓.

Both axes have labels with units of measurement ✓.

The points are accurately plotted but joined by curved lines ✗.

The lines for 4°C and 18°C are identified ✓.

The candidate scores 4 marks.

(ii) There is a greater concentration of potassium chloride inside than outside, so it can't move in by diffusion ✓.

🖉 This is correct, for 1 mark.

(iii) At higher temperature, the rate of respiration is greater ✓, so more ATP is available for the action of the carriers in active transport ✓.

🖉 The candidate has correctly identified that active transport requires ATP, and that the rate of respiration is dependent on the temperature. The candidate scores both marks.

(iv) From 0 mM to 20 mM: there are more ions available for transport ✗.

From 20 to 40 mM: the carrier proteins are functioning at their maximum rate ✓.

🖉 The answer to the first part should have specified the frequency of attachment of the ions to the carrier proteins. The answer to the second part is correct and earns 1 mark.

(v) Rubidium ions inhibit respiration ✗.

🖉 This would reduce the rate of K^+ ion uptake, but the answer does not use the information given in the question — that rubidium ions have similar properties to potassium ions. The candidate fails to score.

🖉 **Overall, Candidate A scores 8 marks and Candidate B scores 12.**

Question 7

(a) The polymerase chain reaction (**PCR**) enables many copies of **DNA** to be made from a small sample. The diagram below summarises the procedure.

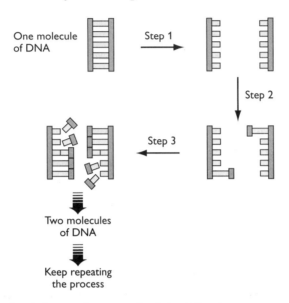

One molecule of DNA

Step 1

Step 2

Step 3

Two molecules of DNA

Keep repeating the process

(i) Explain what is happening at each of the following steps:
Step 1
Step 2
Step 3   (3 marks)

(ii) At the end of the first cycle, there are two molecules of **DNA**.
How many molecules will there be at the end of five cycles?   (1 mark)

(b) **DNA** fingerprinting (profiling) is based on the genetic uniqueness of each individual (identical twins excepted). Particular enzymes are used to cut human **DNA** at specific sites to produce fragments of different length. The **DNA** fragments are separated by gel electrophoresis (which distinguishes the fragments on the basis of size) to produce a series of bands. Each person's **DNA** produces a unique set of bands.

(i) Name the type of enzyme used to cut **DNA**, and explain how the site is recognised by the enzyme.   (2 marks)

(ii) Suggest why this type of enzyme cuts the **DNA** to produce fragments of different lengths.   (1 mark)

The results of **DNA** fingerprinting can provide strong evidence in cases of disputed parenthood. The diagram below represents **DNA** fingerprints of a mother (**M**) and her child (**C**), and those of three men (**X**, **Y** and **Z**).

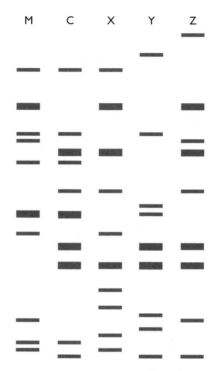

(iii) **How many DNA fingerprints do the mother and child have in common?** (1 mark)

(iv) **Which individual (X, Y or Z) is the father of the child? Explain your answer.** (2 marks)

(c) **A DNA fragment, consisting of 24 base pairs, was analysed for the number of different bases on each strand. The table below shows some of the results. Determine the missing values and complete the table.**

	A	G	T	C
Strand 1	7	8		
Strand 2	4			

(2 marks)

Total: 12 marks

Candidates' answers to Question 7

Candidate A

(a) (i) Step 1: the strands are separated ✗.

Step 2: DNA primers are added to ensure that the two strands don't immediately rejoin ✓.

Step 3: more nucleotides are added and attach to the DNA strands under the control of DNA polymerase ✓.

🖉 The answer to step 1 does not provide sufficient detail, i.e. that heating to 95°C breaks the hydrogen bonds between the two strands. The answers to steps 2 and 3 are correct. The candidate scores 2 marks.

(ii) 32 ✓

🖉 This is correct, for 1 mark.

(b) (i) Restriction endonuclease ✓ recognises the highly repetitive 4-base sequences ✗ in the non-coding sections of DNA.

🖉 The enzyme name is correct, but the candidate does not understand about the recognition sites. There is confusion with the microsatellite repeat sequences (MRSs) used in DNA profiling. The candidate scores 1 mark for restriction endonuclease.

(ii) The length of the fragments depends on the number of repeat sequences ✗.

🖉 The confusion with MRSs is continued. The candidate fails to score.

(iii) 5 ✓

🖉 This is correct, for 1 mark.

(iv) Z ✓, since only Z produces those fragments not supplied by the mother ✓.

🖉 Individual Z is correctly identified and the explanation is excellent. The candidate scores both marks.

(c) Strand 1: T = 7, C = 8 ✗
Strand 2: G = 5, T = 4, C = 5 ✗

🖉 The candidate has equal numbers of A and T bases and of G and C within each strand, rather than *between* the strands (see the answers provided by Candidate B). Candidate A fails to score.

Candidate B

(a) (i) Step 1: the DNA is heated to 95°C to break the hydrogen bonds ✓.

Step 2: DNA primers anneal to the end of each strand to initiate replication ✓.

Step 3: the mixture is heated again and a heat-sensitive DNA polymerase catalyses the addition of complementary nucleotides ✓.

🖉 All the steps are described fully. The candidate scores all 3 marks.

(ii) 64 ✗

🖉 This is incorrect — the candidate has multiplied by 2 one time too many.

(b) (i) Restriction endonuclease ✓ recognises a specific sequence of bases ✓.

 The enzyme and its recognition site are identified correctly, for 2 marks.

(ii) The correct base sequence fits into the active site of the enzyme ✗.

 This does not explain the different number of fragment lengths, i.e. that the specific base sequence occurs at irregular intervals along the DNA. No mark can be awarded.

(iii) 5 ✓

 This is correct, for 1 mark.

(iv) Z ✓, because this individual possesses the most bands in common with the child ✗.

 individual Z is identified correctly, for 1 mark. However, the reasoning is incorrect. It is not just a matter of the number of common bands. The father must provide those fragments (bands) not inherited from the mother.

(c) Strand 1: T = 4, C = 5 ✓

Strand 2: G = 5, T = 7, C = 8 ✓

 This is correct. The candidate has taken base pairing between the two strands into account and that the total number of bases in each strand is 24. The candidate scores both marks.

 Overall, Candidate A scores 7 marks and Candidate B scores 9.

Section A total for Candidate A: 33 marks out of 60

Section A total for Candidate B: 48 marks out of 60

Section B

Quality of written communication is awarded a maximum of 2 marks in this section.   (2 marks)

Question 8

Give an account of the structure and function of the following polysaccharides:
- **starch**
- **glycogen**
- **cellulose**   (13 marks)

Total: 15 marks

Candidates' answers to Question 8

Candidate A

Starch is usually found in plants. It is an energy store in plants ✓ and its glucose isomers are composed from α-glucose ✓. It is a branched structure and has no hydrogen bonds present. Glycosidic bonds are present.

> ✎ Two marking points are given. Other information is not sufficiently detailed — for example, starch cannot be described as branched when branching only occurs in its amylopectin component. The candidate gains 2 marks.

Glycogen is found in the liver and muscles ✓ of animals. It is the energy store ✓ and its glucose isomers are composed of α-glucose and β-glucose. It is a branched structure and also has glycosidic bonds present.

> ✎ Two appropriate points are provided. It is wrong to say the glycogen is a polymer of α-glucose and β-glucose — it is a polymer of α-glucose. Detail is also lacking — branching should be explained as due to 1,6-glycosidic bonds to be worthy of a mark. The candidate gains 2 marks.

Cellulose is found in the cell walls in plants ✓. It has many functions, which include a food source, a structural component of the cell wall and for tensile strength. It is composed of β-glucose ✓ and glycosidic bonds are present.

> ✎ There are two points worthy of marks. Other phrases lack detail – for example, tensile strength is not explained and is not by itself sufficient to earn a mark. The candidate scores 2 marks.

> ✎ The account has a reasonable structure but connections between the points are not sufficiently well established. 1 mark is awarded for quality of written communication.

Candidate B

Starch is the energy store in plants ✓ and is made up of amylose and amylopectin ✓. Both are polymers of α-glucose ✓. They are helical molecules and so compact ✓ while amylopectin is a branched due to 1,6 glycosidic bonds ✓.

> 🖉 Five marking points are given, for 5 marks.

Glycogen is the energy store in animals ✓ particularly in the liver and muscle tissue ✓. It is also a polymer of α-glucose ✓ and is similar to amylopectin in that it is branched due to the presence of 1,6-glycosidic bonds ✓. The many terminal ends mean that it can be rapidly hydrolysed ✓.

> 🖉 Five appropriate points are provided, for 5 marks.

Cellulose is the structural polysaccharide found in the plant cell wall ✓. It is a polymer of β-glucose ✓ and so forms straight chains ✓. The cellulose molecules hydrogen bond together ✓ forming microfibrils ✓ of high tensile strength ✓.

> 🖉 Here, there are six points worthy of marks. However, the candidate can only score a maximum of 3 marks because 10 marks have been awarded already.

> 🖉 The account is well-structured and the statements are well-linked throughout. 2 marks are awarded for quality of written communication.

> 🖉 **Overall, Candidate A scores 7 marks and Candidate B scores 15.**

Section B total for Candidate A: 7 marks out of 15

Section B total for Candidate B: 15 marks out of 15

Paper total for Candidate A: 40 marks out of 75

Paper total for Candidate B: 63 marks out of 75

Exemplar paper

Section A

Question 1

Read through the following passage on the cell cycle, and write the most appropriate words in the blank spaces to complete the account.

Actively dividing eukaryotic cells go through a process called the cell cycle. During .... there are intense periods of growth within which there is a .... phase when **DNA** replication occurs. In the nuclear division, mitosis, four phases are recognised. In the first of these, the chromosomes shorten and thicken, and each is seen to consist of a pair of chromatids. In the next phase, known as .... , the chromosomes assemble across the equatorial plane. This is followed by a phase during which two nuclei form. .... follows nuclear division and involves the division of cytoplasm and the formation of two daughter cells. Two nuclear divisions occur during meiosis so that the chromosome number is.......... .   (4 marks)

Total: 4 marks

Candidates' answers to Question 1

Candidate A
Interphase ✓; synthesis ✓; metaphase ✓; anaphase ✗; 23 ✗

> The first three responses are correct and earn 2 marks. Cytokinesis follows nuclear division, so anaphase is incorrect. The candidate has not read the passage carefully enough. Meiosis produces haploid cells — 23 is correct only for human cells and so cannot be awarded a mark.

Candidate B
Interphase ✓; synthesis ✓; metaphase ✓; cytokinesis ✓; hapliod ✓

> All five answers are correct and earn the full 4 marks. Even though the spelling of haploid is incorrect it is not sufficiently poor as to prevent recognition.

> **Overall, Candidate A scores 2 marks and Candidate B scores 4.**

Question 2

The diagram below represents three adjacent plant cells.

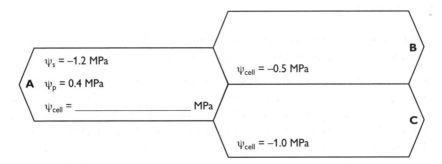

(a) Calculate the water potential of cell **A**. (1 mark)

(b) Show, by drawing arrows on the diagram, the direction of water movement between the three cells. (2 marks)

(c) With reference to water potential, explain why water moves in the direction that you indicated. (1 mark)

(d) Explain what would happen to the water potential of cell **A** if sugars were produced in the cell. (1 mark)

Total: 5 marks

Candidates' answers to Question 2

Candidate A

(a) For cell A, $\psi_{cell} = -0.8$ MPa ✓

> @ This is correct, for 1 mark. If the negative sign had been omitted, the mark would not have been awarded.

(b) Arrows from A to C ✓, B to A ✓ and B to C ✓

> @ The arrows are all correct, for 2 marks.

(c) Water moves from high concentration to low concentration ✗.

> @ This is incorrect, since the term water potential, requested in the question, is not used. Always take care when reading questions. Further, the term concentration is too vague because it could refer to either water or solute. It is preferable to think about the amount of 'free' water, i.e. the water molecules that are not attracted to solute molecules. The candidate fails to score.

(d) Cell A would have a lower ψ_{cell} as the sugar would decrease the ψ_s of the cell (as $\psi_{cell} = \psi_s + \psi_p$) ✓

🖉 This full answer earns the mark.

Candidate B

(a) $\psi_{cell} = -0.8\,MPa$ ✓

🖉 This is correct, for 1 mark.

(b) Arrows from A to C ✓, B to A ✓ and B to C ✓

🖉 The arrow directions are all correct, for 2 marks.

(c) Water moves from an area of high ψ to an area of low ψ (more negative) ✓

🖉 The candidate understands that -1 is lower than both -0.8 and -0.5, and earns the mark.

(d) ψ_{cell} would increase as the presence of more solutes means that ψ_s would become higher ✗.

🖉 The candidate has forgotten that the addition of solutes decreases the solute potential and so consequently ψ_{cell} becomes lower. The candidate fails to score.

🖉 **Overall, both candidates score 4 marks.**

Question 3

(a) Four test tubes each contained a solution of a single carbohydrate: fructose, glucose, sucrose or starch.

The test tubes were labelled **A** to **D** and, in order to identify the carbohydrate present in each tube, a series of tests was carried out.

(1) Samples from all four tubes were tested with iodine solution. Tube **B** tested positive.

(2) Samples from the remaining three tubes were tested with clinistix. Tube **A** tested positive.

(3) Samples from the remaining two tubes were tested using Benedict's test. Tube **D** tested positive.

The remaining tube can be identified by elimination.

(i) Using the test results shown above, identify the carbohydrate in each of tubes **A** to **D**, and complete the table below.

Tube	Carbohydrate
A	
B	
C	
D	

(3 marks)

(ii) Complete the table below to indicate a positive result for each test.

Test	Positive result
Iodine	
Clinistix	
Benedict's	

(2 marks)

(b) Sugars, particularly sucrose, are added to soft drinks as sweeteners. However, not all sugars impart the same level of sweetness. Glucose has only 75% of the sweetness of sucrose but fructose is twice as sweet as glucose. Determine how much fructose must be added to replace 30 g of sucrose. Show your working. (2 marks)

Total: 7 marks

Candidates' answers to Question 3

Candidate A

(a) (i) A — fructose ✗
     B — starch ✓
     C — glucose ✗
     D — sucrose ✗

> Only the iodine test for starch has been identified correctly. The tests for the sugars are not well understood. Clinistix is a specific test for glucose (test A), while sucrose is not a reducing sugar and, therefore does not give a positive result with Benedict's solution (test D). The candidate gains only 1 mark.

(ii) Iodine — blue–black ✓
    Clinistix — purple ✓
    Benedict's — orange–red ✓

> All three responses are correct, for 2 marks.

(b) 75% of 30 g = 22.5 g ✗
   22.5 ÷ 2 ✓ = 11.25 g

> If glucose has 75% of the sweetness of sucrose, then more of it is required to give the same level of sweetness. So 40 g of glucose equates to 30 g of sucrose which equates to 20 g (40 ÷ 2) of fructose, since fructose is twice as sweet as glucose. The candidate has not got the first relationship, but understands to divide by 2 for the link between glucose and fructose and so earns 1 mark.

Candidate B

(a) (i) A — glucose ✓
     B — starch ✓
     C — sucrose ✓
     D — fructose ✓

> All answers are correct, for 3 marks.

(ii) Iodine — blue–black ✓
    Clinistix — colourless ✗
    Benedict's — brick-red ✓

> The end-point colour for the Clinistix test is purple, although a range of colours from purple to blue is acceptable. There is also an acceptable range of colours for a positive Benedict's test. Two correct responses gain 1 mark.

(b) sweetness of fructose = 150% (75% × 2) sweetness of sucrose ✓
   so 30 g ÷ 150% = 20 g ✓

> The candidate has worked out the relationships well and seems at ease with numerical problems. Both marks are awarded.

> **Overall, Candidate A scores 4 marks and Candidate B scores 6.**

Question 4

(a) The diagram below shows the structure of a typical prokaryotic cell (a bacterium).

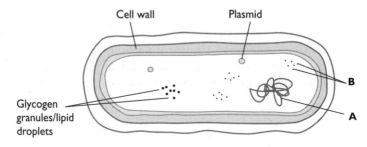

(i) Identify the structures labelled **A** and **B**. (2 marks)

(ii) Suggest one function of each of the following structures found in the bacterial cell:
- glycogen granules
- cell wall (2 marks)

(b) Mitochondria are absent from prokaryotic cells but are present in eukaryotic cells. Describe the structure and function of mitochondria. (2 marks)

(c) Viruses occur in a variety of forms, including bacteriophages (phages) and the human immunodeficiency virus (HIV).

(i) List the molecules of which all viruses are composed. (2 marks)

(ii) Explain why viruses are not regarded as living cells. (1 mark)

Total: 9 marks

Candidates' answers to Question 4

Candidate A
(a) (i) A — DNA ✓

B — ribosomes ✓

☑ Both answers are correct, for 2 marks.

(ii) Glycogen granules — to store food ✗

Cell wall — to provide support ✓

☑ Glycogen is a store of glucose (energy) — suggesting that it is a store of food is too vague. The cell wall does provide support and a mark is awarded. However, Candidate B provides a better answer.

(b) Mitochondria have an envelope of two membranes, the inner being folded to form cristae ✓. They are used to produce energy ✗.

> 🖉 The structure is sufficiently well described. However, it is not precise enough to say that mitochondria produce energy. They produce ATP through aerobic respiration. The candidate gains 1 mark.

(c) (i) Protein ✓, DNA ✗

> 🖉 All viruses contain protein, but not all contain DNA. Some, such as HIV, contain RNA. The candidate earns 1 mark for protein.

(ii) They need a host to do anything ✗.

> 🖉 This could be describing any parasite. The candidate fails to score.

Candidate B

(a) (i) A — DNA ✓; B (blank) ✗

> 🖉 A is correct, for 1 mark. B can only really be ribosomes — they cannot, for example, be vesicles since a prokaryotic cell has no membrane-bound organelles.

(ii) Glycogen granules — store glucose for respiration ✓

Cell wall — a structural role in preventing osmotic bursting of the cell ✓

> 🖉 These full answers earn both marks.

(b) Mitochondria are surrounded by two membranes with the inner membrane folded to form cristae ✓. They contain a matrix with small ribosomes and a circular DNA molecule. In aerobic respiration much of the cell's ATP is produced ✓.

> 🖉 This excellent answer contains more than is required for the 2 marks.

(c) (i) Protein ✓, nucleic acid ✓

> 🖉 Both answers are correct, for 2 marks.

(ii) Viruses have no cellular structure or metabolic activity, and can only replicate using the metabolism of a host cell ✓.

> 🖉 This excellent answer earns the mark.

> 🖉 **Overall, Candidate A scores 5 marks and Candidate B scores 8.**

Question 5

The photograph below is a photomicrograph of a transverse section through part of the wall of the small intestine (ileum).

Draw a block diagram to show the tissue layers in the ileum, as shown in the photograph. Label the drawing to identify at least five structures.

(9 marks)

Total: 9 marks

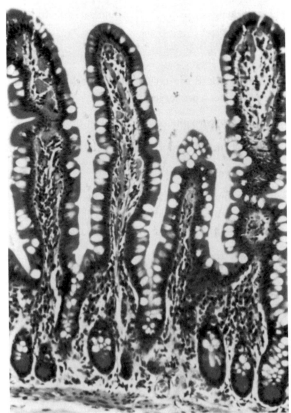

Biophoto Associates/SPL

Candidates' answers to Question 5

Candidate A

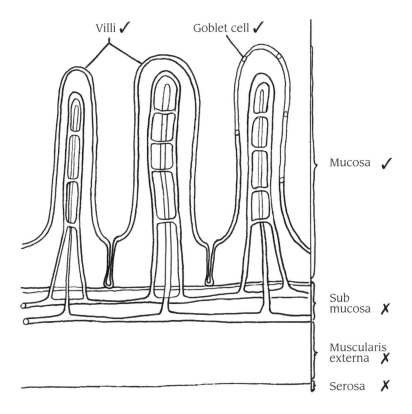

Villi ✓   Goblet cell ✓

Mucosa ✓

Sub mucosa ✗

Muscularis externa ✗

Serosa ✗

✎ This is a well-learned textbook diagram. It is a block diagram showing tissue layers ✓ and has a degree of completeness in showing the tissues obvious in the photograph ✓. However, it is not an accurate representation of the photograph ✗ and it lacks the proportionality of the features shown ✗. The lines drawn are smooth and continuous, not sketchy ✓. The candidate earns 3 marks out of 5 for drawing skills.

✎ Five labels are required for 4 marks etc. The candidate has three features correct, so gains 2 marks. The layers submucosa, muscularis externa and serosa are not included in the photograph and so cannot be awarded marks.

Candidate B

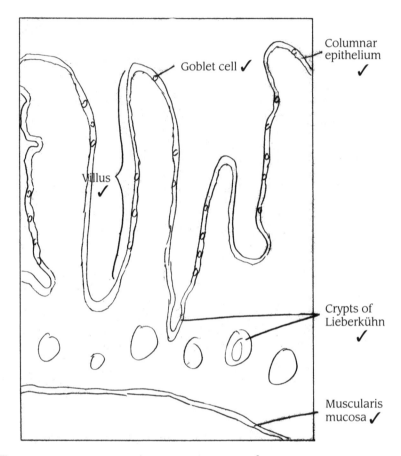

📝 This is a block drawing showing tissue layers ✓ that illustrates all the obvious features ✓. It is a fair attempt to draw the photograph ✓ and the proportionality of features is sufficiently accurate ✓. However, the lines tend to be sketchy in places, and circular structures, such as those for the goblet cells, are incomplete ✗. The candidate earns 4 marks out of 5 for drawing skills.

📝 Five labels are correctly identified, for 4 marks.

📝 **Overall, Candidate A scores 5 marks and Candidate B scores 8.**

Question 6

(a) Enzymes are globular proteins that catalyse metabolic reactions. The diagram below shows the arrangement of the amino acids in an enzyme to which substrates have attached.

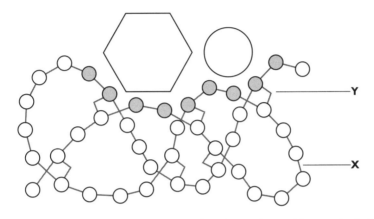

(i) Name the part of the enzyme in which the amino acids are shaded grey. (1 mark)

(ii) Use the diagram and your understanding of enzyme action to explain how an enzyme acts as a catalyst. (3 marks)

(iii) Identify the types of bond labelled **X** and **Y**. (2 marks)

(b) The graph below shows the effect of changes in temperature on the activity of the enzyme sucrase, which catalyses the hydrolysis of sucrose.

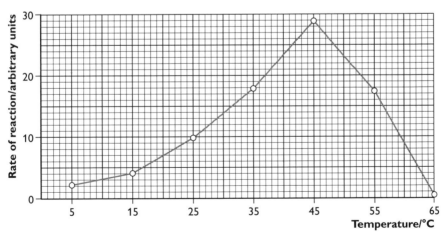

Explain the effect of a change in temperature from 5°C to 35°C on the activity of sucrase. (2 marks)

(c) The graph below shows the effect of varying the concentration of sucrose on the activity of the enzyme, sucrase.

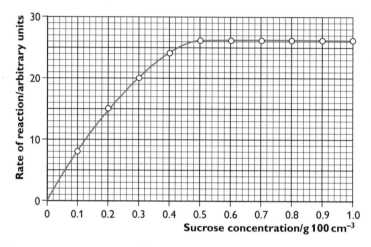

(i) Identify the trends evident in the graph. (2 marks)

(ii) Explain the trends identified. (2 marks)

(d) The chemical mercaptoethanol breaks (reduces) disulphide links. If the mercaptoethanol is then removed and the enzyme exposed to an oxidising environment, the disulphide links are re-formed.

(i) Explain the effect that mercaptoethanol would have on the activity of an enzyme. (1 mark)

(ii) Explain the effect that heating to temperatures above 45°C would have on the activity of an enzyme such as sucrase? Suggest how this differs from the effect of mercaptoethanol. (2 marks)

Total: 15 marks

Candidates' answers to Question 6

Candidate A
(a) (i) Active site ✓

  🖉 This is correct, for 1 mark.

(ii) They speed up the rate of reaction without being altered themselves. Enzymes bind molecules called substrates and promote the reaction that changes the substrate to products.

  🖉 This answer is not precise enough. It contains nothing about the complementary nature of the active site and substrate or about effectively lowering the activation energy. The candidate fails to score.

(iii) X — hydrogen bond ✗

Y —peptide bond ✗

🖉 The candidate has not looked carefully enough at the diagram: peptide bonds link amino acids into a chain (so bond X); disulphide, ionic or hydrogen bonds link amino acids into folds (so bond Y). The candidate fails to score.

(b) As temperature is increased from 5°C to 35°C the rate of reaction increases. This is due to the increased kinetic energy of the molecules ✓.

🖉 The first sentence *describes* the trend rather than *explaining* it. The second sentence does not go far enough – for example, to note the increased chances of collision and of forming enzyme–substrate complexes. The candidate scores only 1 mark.

(c) (i) As sucrose concentration increases from 0 to 0.5, the rate of reaction increases ✓. After this, there is no effect.

🖉 The first point is correct, for 1 mark. In the second point, 'after this' suggests a time element, which is not present. Furthermore, it is not that there is no effect — the rate of reaction is high. The point is that after $0.5\,g\,100\,cm^{-3}$, there is *no further increase* in the rate of reaction.

(ii) As more sucrose is added there are more collisions between enzyme and substrate molecules✓ until the enzyme concentration becomes the limiting factor ✓.

🖉 This good answer, with correct reference to both trends, scores 2 marks.

(d) (i) It would alter the active site leaving the enzyme inactive ✓.

🖉 This is correct, for 1 mark.

(ii) Above 45°C, bonds within the enzyme break and the enzyme becomes inactive ✓.

🖉 The effect of heating is explained, but there is no attempt to suggest how the heating effect differs from the effect of mercaptoethanol. Therefore, the candidate scores only 1 mark.

Candidate B

(a) (i) Active site ✓

🖉 This is correct, for 1 mark.

(ii) The reaction takes place on the active site, which has a complementary shape to that of the substrate molecules ✓. Enzymes effectively lower the activation energy needed for the reaction ✓ by orientating the substrates in such a way as to facilitate bonding between them ✓.

🖉 This excellent answer gains all 3 marks.

(iii) X — hydrogen bonds ✗
   Y — disulphide bonds ✓

📝 X represents a peptide bond, so the candidate scores only 1 mark.

(b) Increasing the temperature increases the kinetic energy of enzyme and substrate molecules ✓. Therefore, they collide more frequently, which increases the formation of enzyme–substrate complexes ✓.

📝 This full answer earns 2 marks.

(c) (i) The rate of reaction increases as substrate concentration increases, up to $0.5\,g\,100\,cm^{-3}$ ✓. Above this concentration of sucrose, the rate of reaction levels off ✓.

📝 Both trends are described clearly, for 2 marks.

(ii) As the sucrose concentration is increased, there are more collisions between the enzymes' active sites and the substrate molecules so more enzyme–substrate complexes are formed ✓.

📝 This is correct as far as it goes, for 1 mark. However, the candidate has forgotten to explain *why* the graph levels off — at high concentrations of substrate the active sites of the enzyme molecules are saturated.

(d) (i) It would temporarily inhibit the enzyme, as it loses its active site, though when mercaptoethanol is removed the enzyme will become active again ✓.

📝 This is correct, for 1 mark.

(ii) At high temperature bonds holding the enzyme in a precise shape are broken and the enzyme is denatured ✓. This is permanent damage, while the effect of mercaptoethanol is not ✓.

📝 This correct answer earns 2 marks.

📝 **Overall, Candidate A scores 7 marks and Candidate B scores 13.**

Question 7

(a) The diagram below shows part of a **DNA** molecule in the process of replication.

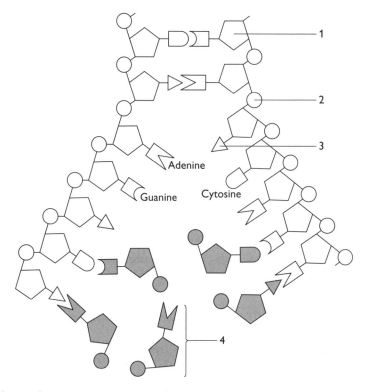

(i)   Name the structures labelled 1 to 4.   (4 marks)

(ii) Name the enzyme which catalyses the attachment of structure 4
      to the **DNA** strand.   (1 mark)

(iii) What term is used to describe the mechanism by which **DNA**
      replicates?   (1 mark)

(b) The **DNA** of each person is different. Differences in the **DNA** may
    be analysed using restriction endonuclease enzymes that cut **DNA**,
    producing fragments of different lengths.

(i)   Explain why a specific restriction endonuclease enzyme cuts two
      DNA samples to produce a different number of fragment lengths. (2 marks)

(ii) What term is used to describe the various lengths of **DNA** from different
      people as revealed by analysis using endonuclease enzymes?   (1 mark)

(iii) A particular DNA fragment may be located using a DNA probe. Explain what is meant by DNA probe. (2 marks)

Total: 11 marks

Candidates' answers to Question 7

Candidate A

(a) (i) 1 — pentose sugar ✗
2 — phosphate ✓
3 — organic base ✗
4 — nucleotide ✓

🖉 Structure 1 is deoxyribose; pentose is not precise enough to gain the mark. Structure 3 is an organic base, but should be identified as thymine (will pair with adenine opposite). The candidate earns 2 marks for naming structures 2 and 4 correctly.

(ii) DNA polymerase ✓

🖉 This is correct, for 1 mark.

(iii) The S phase ✗

🖉 The S phase is the stage in the cell cycle during which DNA replication takes place. The mechanism is known as semi-conservative replication. The candidate fails to score.

(b) (i) This enzyme cuts the DNA at sites that have a specific base sequence ✓.

🖉 The candidate understands the recognition sites of endonuclease enzymes. However, as there is no mention of why a different number of fragment lengths are produced, the candidate earns only 1 mark.

(ii) Restriction fragment length polymorphism ✓

🖉 This is correct, for 1 mark.

(iii) A probe is a short length of single-stranded DNA ✓ with a known nucleotide (base) sequence ✓.

🖉 This excellent answer earns both marks.

Candidate B

(a) (i) 1 — pentose (5C) sugar ✗
2 —phosphate ✓
3 — nitrogen-containing base ✗
4 — nucleotide ✓

🖉 Deoxyribose is a more accurate answer for structure 1, as is thymine for structure 3. The candidate scores 2 marks.

(ii) DNA helicase ✗

🖉 DNA helicase is the enzyme that breaks the hydrogen bonds to 'unzip' the two strands of DNA. The answer is DNA polymerase. The candidate fails to score.

(iii) Semi-conservative replication ✓

🖉 This is correct, for 1 mark.

(b) (i) The endonuclease enzyme only cuts at a specific base sequence ✓ and the DNA of different individuals possesses a different number of these recognition sites ✓.

🖉 This excellent answer earns both marks.

(ii) DNA fingerprinting ✗

🖉 DNA profiling or fingerprinting uses a variety of techniques to determine the differences in the DNA of different individuals. The specific answer required here is restriction fragment length polymorphism. No mark can be awarded.

(iii) A short length of single-stranded DNA ✓ with a specific base sequence ✓ that will attach to a specific section of DNA.

🖉 This excellent answer earns both marks.

🖉 **Overall, both candidates score 7 marks out of 11.**

Section A total for Candidate A: 34 marks out of 60

Section A total for Candidate B: 50 marks out of 60

Section B

Quality of written communication is awarded a maximum of 2 marks in this section. (2 marks)

Question 8

(a) Describe the structure of the cell-surface (plasma) membrane. (7 marks)

(b) Explain how membrane structure determines how molecules pass through the membrane. (6 marks)

Total: 15 marks

Candidates' answers to Question 8

Candidate A

(a) The cell membrane is composed of proteins ✓ floating in a fluid bilayer of lipid. This structure is called the fluid-mosaic model ✓. It is made of two layers of phospholipids ✓ that have heads and tails. The tails of these two layers face each other with the heads facing in opposite directions. The tails are non-polar ✓ and the heads are polar✓. Proteins consist of long chains of amino acids. Some of these are polar and some aren't. Some rest on the surface of the bilayer while others go right through ✓. The ones on the surface generally have carbohydrate attached producing glycoproteins ✓.

> ✍ Seven appropriate points are given. In some cases, however, the marks are only just arrived at. 'Bilayer of lipid' is not sufficient to earn a mark, though a following sentence describes 'two layers of phospholipids'. Marks are awarded over two sentences for polar heads facing outermost and non-polar tails facing innermost. The comment about some of the amino acids in the proteins being polar and some being non-polar is not sufficient to earn a mark. The candidate should have stated that the amino acids in contact with the lipid layer are non-polar. The candidate scores 7 marks.

(b) Proteins and glycoproteins may act as carriers. Other proteins act as enzymes. Cells have to be recognised by antibodies and hormones because of proteins on the surface. Water is able to pass between the phospholipid molecules of the bilayer ✓. Water-soluble substances also use this route. Glucose, which is polar, relies on carrier proteins ✓. Facilitated diffusion takes substances against the concentration gradient.

> ✍ Two appropriate points are provided. Some points are simply wrong, e.g. lipid-soluble substances (not water-soluble) pass directly through the bilayer while facilitated diffusion moves substances down (not against) the concentration

gradient. Other points are not relevant: proteins acting as enzymes or hormone receptors have nothing to do with movement across the membrane. The candidate scores 2 marks.

🖉 The candidate expresses ideas clearly, though not always fluently, and the account has sometimes strayed from the point. 1 mark is awarded for quality of written communication.

Candidate B

(a) The cell surface membrane consists of a bilayer of phospholipids ✓. The heads of phospholipids are polar and hydrophilic, so they are arranged outermost in water ✓. The tails of the phospholipids are non-polar and hydrophobic, so they remain in contact with each other ✓. There are proteins ✓ interspersed among the fluid phospholipids, so the structure is described as a fluid-mosaic model ✓. There are intrinsic proteins that are embedded within the bilayer and there are extrinsic proteins ✓. Some proteins have carbohydrate attached ✓ and this acts as a cell recognition feature. Cholesterol is also present among the phospholipids and stabilises the fluidity of the membrane ✓, especially when the temperature changes.

🖉 Eight marking points are given, so the candidate scores all 7 marks.

(b) Small gaps in the phospholipids allow entry of small, polar molecules such as water, CO_2 and O_2 by simple diffusion ✓. Molecules that are too large and polar must pass through protein channels ✓. Each has a site for a specific molecule so there are many different types of protein channel for different types of molecule ✓. Carrier proteins can also change shape — a molecule binds to the protein, it changes shape and the molecule is released inside the cell ✓. Facilitated diffusion is passive ✓, requiring no energy input. Active transport goes against the concentration gradient and requires energy in the form of ATP ✓. Where active transport takes place, there is a large number of mitochondria.

🖉 Six appropriate points are provided, for 6 marks.

🖉 This is a well-structured account with the ideas expressed fluently. The relationship between membrane structure and how substances pass across the membrane is made clearly. 2 marks are awarded for quality of written communication.

🖉 **Overall, Candidate A scores 10 marks and Candidate B scores 15.**

Section B total for Candidate A: 10 marks out of 15

Section B total for Candidate B: 15 marks out of 15

Paper total for Candidate A: 44 marks out of 75

Paper total for Candidate B: 65 marks out of 75